Magnetic Resonance in Studying Natural and Synthetic Materials

Authored by

Victor V. Rodin

Institute of Organic Chemistry, Johannes Kepler University Linz, 4040 Linz, Austria

Bentham Science Publishers Ltd.
Executive Suite Y - 2
PO Box 7917, Saif Zone
Sharjah, U.A.E.
Email: subscriptions@benthamscience.org

CONTENTS

FOREWORD 1

"Water, Water, Everywhere" begins a famous quatrain by the English poet Samuel Taylor Coleridge. Indeed, there is no chemical compound more ubiquitous or more crucial to all aspects of life on Earth. Most familiar is water in its liquid form, a material that is a near-universal solvent. Sea water, a solution of inorganic salts in water, represents more than 95% of the water on the surface of the planet. While water molecules are essential to life, many aspects of their involvement in biological structures and processes are not fully understood.

Formed from a single oxygen atom and two hydrogens, the physical and chemical properties of water are often anomalous if the hydrogen compounds of the elements surrounding oxygen in the periodic table are considered. For example, water (H_2O) is a liquid at room temperature and atmospheric pressure while the compounds NH_3, HF and H_2S are gases under the same conditions. The anomalous properties of water result from intermolecular dipolar and hydrogen-bonded interactions that are peculiarly potent in pure water and account for this liquid's ability to function so effectively as a solvent. The same types of interactions often lead to water molecules being found within organized three-dimensional structures of living and inanimate materials where they have specific structural roles that go beyond those of a mere solvent.

Many experimental methods have been applied to the study of water as a solvent or as a part of an assembled structure. Given that water contains hydrogen atoms with their spin ½ nuclei, it is no surprise that proton nuclear magnetic resonance (^{1}H NMR) has been at the forefront of efforts to understand the chemistry of water wherever it is found. Proton NMR can provide information about the stability of organized structures and indications of the time rate(s) of change of these. Such changes might include rearrangement of three-dimensional aspects (such as conformational motions) or change in relative positions through rotational and translational diffusion.

In this volume, Dr. Rodin presents results from his laboratory and those of others that exemplify information that can be obtained about water molecules contained within organized but non-crystalline materials using proton NMR. These efforts appropriately have employed various relaxation, multiple quantum filtered, translational diffusion and imaging experiments. The systems examined range from synthetic (polyacrylates) and natural polymers (collagen and silk) to intact wood and inorganic cements. These reports nicely demonstrate to the reader the current state of the art in applying these powerful NMR experiments to the systems mentioned.

J. T. Gerig, Professor Emeritus
University of California
Santa Barbara
USA

FOREWORD 2

Nuclear Magnetic Resonance (NMR) is an indispensable technique for investigating the structure, functionality and dynamics of molecules. It is used widely in physics, chemistry, biology, medicine and other sciences. In this book, Dr. Victor Rodin describes NMR experiments and techniques for consideration and discussion. This book is a great addition to the analysis of NMR methods which Dr. Victor Rodin applied in studying the translational dynamics of molecules in porous and heterogeneous systems: synthetic polymers and materials, collagens and natural silk, and wood and cement pastes.

Dr. Rodin graduated from the Moscow Institute of Physics and Technology (MIPT-State University), Faculty of Molecular & Chemical Physics. He received a PhD in Biophysics (with specialization in Mathematics & Physics) at the Research Institute of Biophysics (Russian Academy of Sciences). He also earned a PhD in Macromolecular and Colloid Chemistry at the Department of Chemistry of Moscow State University. He has worked in excellent NMR research centers, including the University of California, Santa Barbara, USA; the University of East Anglia; the University of Bristol; the University of Surrey, UK; INRA, Clermont-Ferrand, France; Johannes Kepler University of Linz, Austria. Dr. Rodin has published more than 60 publications in peer-reviewed scientific journals and delivered approximately 70 presentations at international scientific meetings and conferences. He has been a member of the Research Council on Colloid Chemistry at Moscow State University, and a member of the International Society of Magnetic Resonance in Medicine, USA. He has refereed papers for scientific journals including *Colloid Journal*, *Material Science*, *Polymer*, *Food Chemistry*, and a special issue of *Magnetic Resonance in Porous Media*.

Dr. Rodin has considerable experience in the development of different magnetic resonance methods, including new methodologies. His research experience focuses on the development and application of MR methods and analysis to study solutions and heterogeneous materials, including biomaterials and drugs, blood and microbiological suspensions, polymer solutions, gels and films, xenon gas clathrate hydrates, collagen tissues, natural silk, skin, wood, cement etc. He has taught in many universities. Based on his research results he has delivered lectures to students in physics, physical chemistry, mathematics, natural biopolymers and synthetic materials.

Dr. Rodin illustrates the current state of numerous special NMR experiments applied to porous heterogeneous materials. Readers with an interest in NMR will find useful information in this volume. Potential roles of NMR to future applications are also discussed. The examples are taken from real research results. Problems and solutions are also considered.

Syed F. Akber, PhD, DABR
Radiological Physicist
Cleveland/ Lorain, OHIO
USA

FOREWORD 3

NMR has been used extensively to estimate the amount of water as well as water self-diffusion in porous natural as well as man-made polymeric materials. It is well known that water is not simply a medium in which biomolecules happen to exist, interact and carry out their functions. Almost every aspect of biomolecular chemistry and physics is influenced by the properties of the liquid milieu. The changes in hydration dynamics of biomolecules upon interaction with other molecules are critical for its functional applications such as for tissue engineering. The author himself is very well versed in understanding hydration dynamics of macromolecules especially silk and collagen, the widely used biomaterials in tissue engineering. Dr. V.V. Rodin's experience in understanding water diffusivity with other polymeric materials *viz.*, wood and acrylic polymers is also immense. Materials used in cement industry are within his research as well.

The book starts with the basic understanding of NMR principles and concentrates on the description of NMR relaxation and diffusion experiments to understand dynamic properties of bound water in natural polymers and explains in detail the theory behind how such experiments work. The 2D diffusion-diffusion correlation NMR spectroscopy needed to study diffusion anisotropy in wood is introduced step by step, with the emphasis on obtaining a good understanding of how the experiments actually work. Water-polymer interactions and self-diffusion of water in polymer films studied using NMR are also detailed vividly.

Biophysicists and polymer chemists who read this book will be rewarded with fundamental understanding of number of NMR methodologies in studying different heterogeneous materials at a level that will allow them to make use of this versatile spectroscopic method for investigating water dynamics of natural proteins, synthetic polymers and other porous heterogeneous materials.

N. Nishad Fathima
Principal Scientist, PhD
CSIR-Central Leather Research Institute
India

FOREWORD 4

This is a book about studies of water motion by the author and his collaborators for a number of natural polymers, such as wood and silk, for the synthetic polymers, and for some cement pastes. Various advanced NMR relaxation and diffusion measurements were performed. These include T_1-T_2 cross correlation relaxation and diffusion-diffusion correlation spectroscopy with parallel and perpendicular pulsed field gradients. A measurement of self diffusion of a polymer itself is also made.

Victor Rodin, working in magnetic resonance methods at the University of Linz, at Linz, Austria, has had a life long professional interest in the physical properties of natural and synthetic polymers, including collagen, wood, and mineral type materials, especially in regard to the static and dynamic properties of water associated with these materials. Studies have also included water bound in clathrates.

Dr. Rodin has authored or co-authored over 60 articles in English and Russian journals, including a chapter in the 2017 Encyclopedia of Physical Organic Chemistry.

Hopefully, readers with different backgrounds in physics, chemistry, and biology will acquire much useful information from these studies of water-macromolecule and pore-water interactions in organized structures. Not only is the information about the systems themselves of value; but the methods used to acquire the information can be applied to many other systems.

John E. Tanner, Jr., PhD
USA

PREFACE

This book presents an analysis of those nuclear magnetic resonance (NMR) methods which are used in the studying translational dynamics of molecules in different complex systems including synthetic and natural polymers, tissues, and the porous heterogeneous systems of different destination. The results of proton spin-lattice and spin-spin relaxation, cross-relaxation, pulse field gradient (PFG) NMR in studying diffusion properties and dynamics of molecules in polymer systems of different complexity are reported: from polymer solutions to polymer films and biomaterials of natural origin. The book describes a number of NMR methodologies in studying different heterogeneous materials. In addition to these methods, double-quantum-filtered (DQF) NMR technique in a study of slow molecular dynamics and properties of the systems with the anisotropic properties is presented. DQF NMR is working in investigating the systems, in which there is an order. It is effectively applied in the systems with anisotropic motion of molecules. The examples are presented on natural silk, collagens, and materials of construction destination. It is considered also how DQF NMR spectroscopy highlights water in hardening cement pastes.

The apparent translational diffusion coefficients (D_{app}) at two orthogonal directions of applied gradient in oriented fibers of natural polymers/materials were studied and discussed to clarify diffusion and estimate the restricted distance and permeability. The book considered the results of the approaches of one-dimensional and two-dimensional NMR and showed how these methods work in addition to the common methods of single-quantum NMR spectroscopy. The book presents also the data of two-dimensional correlation NMR spectroscopy as the distributions of diffusion coefficients in two orthogonal directions on the systems with anisotropic mobility. Simulations of two-dimensional NMR experiments have been done showing how it leads to the explanation of 2D experimental data on the anisotropy of diffusion coefficients. These 2D NMR methods reveal microscopic local anisotropy by the correlation of the diffusion motion of molecules along either collinear or orthogonal directions of applied pulse gradients of magnetic field.

It is shown that combination of NMR relaxation and diffusion (one- and two-dimensional) techniques is effective experimental methodology which is able to produce valuable information on the dynamic properties and anisotropy in natural silk and collagen fibers with various cross-links, synthetic polymers and porous heterogeneous systems, such as wood and cement.

Both collagen and thread produced by silk worm *Bombyx mori* are used in nano-scaffolds fabrication for tissue engineering applications. One of the most features of new-created scaffolds with stable chemical cross-linking is a control of the mechanical properties which change the tissue hydration and macromolecular content. Changes in tissue hydration, water-fiber interactions, and macromolecular content can be reflected by NMR. With detailing the hydration properties of collagens at different cross-linking level by NMR, a role of water interactions in improving scaffold characteristics for tissue engineering could be clarified. The book considers how NMR highlights the interaction of these natural fibrous materials with water. Natural silk *Bombyx mori* is also known as one of the strong natural biomaterials. *Bombyx mori* silk fibers have two protein-monofilaments embedded in the glue-like sericin coating. The interaction of these natural materials with water leads to a decrease in the length of silk fibers. One chapter of the book shows how natural silk *Bombyx mori* with low water content has been studied by ^{1}H DQF NMR and single-pulse ^{1}H NMR methods. That investigation demonstrated that DQF NMR enables one to probe the anisotropic motion of water in the silk fibers with residual water content.

Polymers are frequently used as barrier coatings in the technology of many materials to prevent the degradation of porous substrates, for example, such as wood. Mostly, a barrier coating should regulate the transport of water. However, there is the fact that water has a solubility even in hydrophobic polymers. So, the questions of water-polymer interaction are essential in many technological stages of material production and could be answered by NMR studies. The penetration of water into polymer films had always increasing attention in material research. Diffusion of water in different polymer films has been extensively investigated in many research groups. Different NMR methods, in particular, NMR relaxometry and NMR spectroscopy have been used to characterize water molecules in the films during the drying process. These approaches have developed the possibilities of identification of different environments for water upon drying the film or in the films swollen in water. The chapters of this book considered also how the diffusivity and distribution of water in polymer films have been investigated using the NMR relaxation and PFG NMR techniques. The contributions of polymer matrix protons, surface water and bound water could be determined from NMR relaxation functions and ^{1}H NMR spectra. The PFG NMR experiments discovered that the echo-attenuation function depends on the diffusion time indicating that water inside the swollen film is trapped in restricted confinement. The published physical models for diffusion of water in polymeric materials have been probed to fit experimental results. As a result, NMR applications were successful in estimating the sizes of pores inside the polymer films. Thus, the studies described in the book established how magnetization decays and the spin-spin relaxation times of water in saturated polymer films could be applied to provide additional information on the water distribution in the porous microstructure.

Wood is one of the most important natural fiber composites. This material is applied in construction area. For fibrous material, the diffusivities of water could be different along the fiber and in the direction of perpendicular to fiber axis. The book considers how to use one- and two-dimensional PFG NMR spectroscopy to explore the anisotropic diffusion of water in wood. Findings herein suggest that it is possible to directly register diffusion anisotropy by one-dimensional PFG NMR as well as to visualize it in 2D maps of wood. These approaches can potentially be used to diagnose and monitor the treatment of wood that involves macromolecular reorganization and associated changes in cross-relaxation and molecular diffusion of water. The methods are applicable for investigating wetting/drying wood. It can be used also in studying modifications of wood technology affecting the diffusion anisotropy of water in wood cells.

Cement pastes are the examples of heterogeneous porous materials with slow water dynamics. Cement reacts with water to form an amorphous paste through the chemical reaction called hydration. It is plastic and soft when newly mixed, strong and durable when hardened in concrete. The character of the concrete is determined by the quality of the paste. To better understand the role of water in mature concrete, it is necessary to understand a diffusion transport of water in cement pastes cured for different ages: from few hrs to few months. The results of one- and two-dimensional PFG NMR diffusometry studies of water in white cement paste aged from 1 day to 1 year show that the NMR PFG method is primarily sensitive to the capillary porosity. Data is fit on the basis of a log-normal pore size distribution with pore size dependent relaxation times. The volume mean capillary pore size is 4.2 μm in mature paste. No evidence is found of capillary pore anisotropy in cement paste.

Hopefully, the NMR results considered would allow one to get more details in description of diffusion anisotropy and properties of different materials in future.

The book is based on the results of own studies and publications. However, all considered data are discussed and compared with literature data presented in this area and used in the context of total directions of the book to highlight main conclusions of NMR studies in polymers, natural biopolymers, and construction materials.

The text of the book is aimed for researchers. However, the level appropriates to a graduate student in physics, mathematics and/ chemistry. The biologist reader may omit mathematical aspects / equations at first reading and find many useful information on NMR applications in biomaterials. The book will be interest and help to both those need to learn additional NMR applications and those needing to refresh their knowledge and extend their own NMR capabilities. Though the book starts at a useful elementary level on basic NMR it goes deeply into the various NMR applications in studying natural biological and synthetic materials. On the other hand, the author mentions / discusses only very briefly certain special matters in NMR which are outside the scope of this book. Hopefully, the material considered is that many readers will find of interest.

Dr. Victor V. Rodin
Institute of Organic Chemistry
Johannes Kepler University Linz
4040 Linz, Austria

CONFLICT OF INTEREST

The author declares no competing interest regarding the publication of this book.

ACKNOWLEDGEMENT

Declared none.

DEDICATION

Dedicated to Prof. Dr. Norbert Müller
as a mark of the respect for his studies in NMR

CHAPTER 1

Basic Principles of NMR and Experimental Techniques

Abstract: This chapter introduces into NMR methods as the most important techniques for studying materials. The behaviour of spin system is considered in crossing magnetic fields from "classical" point of view. Basic equations for the vector of the bulk magnetisation M in outer magnetic field (*Bloch* equations) are discussed in rotating coordinate system. *Spin-lattice (longitudinal)* relaxation and *spin-spin (transverse)* relaxation are considered. The chapter describes the NMR experimental techniques for studying relaxation times T_1, T_2 and cross-relaxation. The role of proton exchange between the water and exchangeable protons of macromolecules is considered. Pulsed field gradient (PFG) NMR techniques are described for application in one- and two-dimensions. Double-quantum-filtered (DQF) NMR spectroscopy is introduced as a technique to study the materials with anisotropic motion of molecules.

Keywords: Apparent diffusion coefficient D_{app}, Double-quantum-filter (DQF) NMR, Fourier transform (FT), Free induction decay (FID), Inverse Laplace transformation (ILT), Nuclear Magnetic Resonance, Pulsed Field Gradient (PFG), Radio frequency (RF), Residual dipolar interaction (RDI), Spin-echo (SE), Spin-lattice (longitudinal) relaxation time T_1, Spin-spin (transverse) relaxation time T_2, Stimulated echo (STE), Two-dimensional diffusion-diffusion correlation NMR spectroscopy (2D DDCOSY).

INTRODUCTION TO NMR

The methods based on the phenomenon of nuclear magnetic resonance (NMR) are effectively used in different fields of physics, chemistry, biology, and applied areas for the investigation of properties and structure of materials [1 - 3]. At present, NMR is the most important technique to study molecular motion and to characterize materials. Many NMR books describe both basic principles of realization of NMR phenomenon in applied techniques and specific topics which provides in-depth state of the reviews on NMR applications in various fields [1 - 7].

Victor V. Rodin

For understanding NMR methods, the magnetic properties of atomic nuclei should be considered at first. Atoms consist of a dense, positively charged nucleus, which is surrounded at a relatively large distance by negatively charged electrons. The particles in atoms (electrons, protons and neutrons) can be imagined as spinning on their axes. In some atoms, *e.g.*, ^{12}C, these spins are paired against each other and the nucleus of the atom has no overall spin. However, in other atoms, such as ^{1}H, ^{13}C, and ^{31}P, the nucleus does possess an overall spin. The nuclei with spin experience NMR phenomenon. There are a large number of nuclei, which do have a nonzero spin angular momentum. NMR is a phenomenon, which occurs when the nuclei of atoms with nonzero spin are placed in a static magnetic field and a second oscillating magnetic field is applied [2 - 5].

The rules for determining the net spin of a nucleus are described as follows. If the number of neutrons and the number of protons are both even, then the nucleus has no spin. If the number of neutrons plus the number of protons is odd, then the nucleus has a half-integer spin, *i.e*, 1/2, 3/2, 5/2. For instance, the nuclei ^{1}H, ^{15}N, ^{13}C, ^{19}F, ^{31}P, and ^{129}Xe have the spin 1/2. The nuclei ^{7}Li, ^{11}B, ^{23}Na, ^{39}K, and ^{131}Xe have the spin 3/2.

If the number of neutrons and the number of protons are both odd, then the nucleus has an integer spin, *i.e*, 1, 2, 3. For example, the nuclei ^{2}H and ^{14}N have integer spin 1, and the nuclei ^{10}B and ^{22}Na have spin 3.

NMR is a quantum phenomenon [4 - 9]. Quantum mechanics tells us that this nuclear spin is characterised by a nuclear spin quantum number, I. According to quantum mechanics, a nucleus of spin I will have $2I + 1$ possible orientations in magnetic field. An interaction of the spin-half nucleus ($I = 1/2$) with a magnetic field results in two energy levels. In the absence of an external magnetic field, these orientations are of equal energy. If a magnetic field is applied, then the energy levels split. Each level is given by a *magnetic quantum number, m*. According to quantum mechanics, m value is restricted to the values from $-I$ to I in integer steps. Thus, for a spin-half nucleus, there are only two values of *magnetic quantum number m*, $+1/2$ and $-1/2$.

There is a tradition in NMR to denote the energy state with $m = +1/2$ as α, which is often described as "spin up" notation. The state with $m = -1/2$ is denoted β and is then described as "spin down" notation. The m values ($+1/2$, $-1/2$) express the parallel and antiparallel orientations of the nuclear spin with respect to the applied magnetic field. Thus, these orientations are described by spin basic functions (α, β). The state with basic function α is the one with the lowest energy [4, 7]. According to quantum mechanics, m value is the *eigenvalue* of the spin operator I_z. When this operator is applied to the one spin basis function $|\psi\rangle$, the result is following: $I_z|\psi\rangle = m|\psi\rangle$. Thus, these two *eigenfunctions* $|\alpha\rangle$ and $|\beta\rangle$ have the properties as shown in eq. (1):

$$I_z|\alpha\rangle = +\frac{1}{2}|\alpha\rangle \qquad I_z|\beta\rangle = -\frac{1}{2}|\beta\rangle \tag{1}$$

Here, eigenvalues are expressed in units of $\hbar = h/2\pi$ [4]. The details of energy levels for two or more spins in molecule can be found elsewhere in literature [1, 3 - 9].

When the nucleus is in a magnetic field, the initial populations of the energy levels are determined by thermodynamics in accordance with the *Boltzmann* distribution. It does mean that in the state of equilibrium the lower energy level will contain slightly more nuclei than the higher level. It is possible to excite the nuclei from the low level into the higher level with an electromagnetic radiation. The frequency of radiation needed is determined by the difference in energy between the energy levels $\Delta E = E_\beta - E_\alpha$.

When positive charged nucleus is spinning, this generates a small magnetic field. Therefore, the nucleus possesses a magnetic moment μ. This magnetic moment is proportional to its spin I, Planck's constant h, and the constant γ, which is called the *gyromagnetic ratio* [4, 5, 9]. γ for nucleus is a ratio of magnetic dipole moment to its angular momentum. γ is a fundamental nuclear constant, which has a different value for every nucleus [2 - 4]. The energy E_m of each energy level m is proportional to the strength of the magnetic field at the nucleus B_0, *magnetic quantum number m* and γh. So, the transition energy ΔE, considered as the difference in energy between levels, will be also proportional to B_0. If the magnetic field B_0 is increased, then ΔE value is also increased. When a nucleus has a relatively large *gyromagnetic ratio*, ΔE is large too. In order to understand how a radiation is absorbed by nucleus in a magnetic field, the behaviour of a charged particle in a magnetic field would be better considered from a "classical" point of view. Let us imagine a nucleus ($I = 1/2$) in a magnetic field. This nucleus is in a base state, *i.e.,* at the lower energy level and its magnetic moment does not oppose the applied field. The nucleus is spinning on its axis. In the presence of a magnetic field, this axis of rotation will *precess* around the magnetic field. The frequency of precession is called the *Larmor* frequency. The *Larmor* frequency is identical to the transition frequency. The potential energy of the precessing nucleus is given by $E = -\mu \cdot B_0 \cdot cos\varphi$, where φ is the angle between the direction of the applied field and the axis of nuclear rotation. The angle of precession, φ, will change when energy is absorbed by the nucleus. For a nucleus with spin of 1/2, the magnetic moment is flipped by absorption of radiation so that it opposes the applied field.

An adsorption of energy results in the higher energy state. Due to the difference in population of levels in the state of equilibrium, there is a macroscopic magnetization *M*, oriented along the direction of magnetic field B_0 (*Z*-axis).

NMR experiments are performed on the nuclei of atoms. The information obtained about the nuclei is used to clarify the chemical environment of many particular nuclei [1 - 5]. NMR researcher normally uses quite often one-dimensional NMR techniques to study chemical structures in frequency domain. In order to determine the structure of complex molecules, in particular, protein structures, two-dimensional techniques are used [4, 7, 9]. Using the time-domain NMR measurements, researchers can study molecular dynamics in solutions, viscous liquids, gels, pastes as well as liquids in porous media [3, 6, 8, 10 - 12]. Liquids filled in porous materials can follow different dynamics at sub-zero temperatures. Different fractions of porous liquids can be discovered [5, 6, 8]. On the base of time-domain NMR experiments, a method for the estimation of the pore sizes and characterization of the materials can be developed.

NMR spectroscopy applies *Fourier transformation* to transfer data between time-domain and frequency-domain. The Fourier transform (FT) is a mathematical technique for converting data from time-domain to data in frequency-domain, and an inverse Fourier transform (IFT) converts data from frequency-domain to time-domain. So, FT is defined as shown in eq. (2), in which $f(t)$ corresponds to the time-domain function; $F(\omega)$ corresponds to the spectrum in the frequency-domain.

$$F(\omega) = \int_{-\infty}^{+\infty} f(t)\exp(-i\omega t)dt = \int_{-\infty}^{+\infty} f(t)\cos(\omega t)dt - \int_{-\infty}^{+\infty} f(t)i\sin(\omega t)dt \qquad \textbf{(2)}$$

$F(\omega)$ in eq. (2) is a complex function consisting of a real and an imaginary parts. It is equally valid to display the spectrum by using either the real part or the imaginary part as the frequency-domain function. In 1D NMR, the spectrum is usually represented by the real part to display the absorption signal [5, 9]. The Fourier transform is relatively simple procedure, *i.e.,* it is possible to implement this on a computer and to generate the frequency-domain signal (spectrum) from the time-domain signal.

According to a "classical" point of view, a behaviour of spin system in crossing magnetic fields (alternative magnetic field $B_1(t) = B_{1m} \cos(\omega t)$ is perpendicular to constant magnetic field B_0) is considered on the base of movement of magnetization vector *M*. Basic equations for the vector of the bulk magnetization *M* in outer magnetic field are the differential *Bloch* equations with taking into account the spin-lattice and spin-spin relaxations [3, 5, 6, 8]. These equations are much easier in a rotating coordinate system $X'Y'Z'$, which rotates with frequency ω around B_0 (Z, Z'-axes) in the direction of nuclear precession [3, 5]. In rotation frame, vector M rotates around magnetic field B_1 (X'-axis). In the absence of magnetic field B_1, magnetization M is along the direction of outer magnetic field B_0. The angle frequency of rotation of vector M around X'-axis is $\omega = \gamma B_1$. The angle θ for rotation of magnetization M during time t_p is presented as $\theta = \gamma B_1 t_p$ [3, 5]. If vector M turns by $\theta = 90°$ during time t_p, then this rotation is called as a 90° or $\pi/2$ pulse (duration time t_p is called the pulse length). When θ=180°, then this rotation is called as a π pulse.

An application of $\pi/2$ pulse to the magnetization M (when originally field B_1 was absent and only field B_0 did work) results in vector M being along the Y'-axis and the intensity of measurable signal (along Y'-axis) has maximal value. During the time (because of relaxation processes) a projection of magnetization vector M to the Y'-axis will decrease. This detected signal is called as a *free induction decay* (FID). Fig. (**1**) shows typical FID (top) and NMR spectrum (bottom) produced in frequency-domain.

SPIN-LATTICE AND SPIN-SPIN NMR RELAXATION

In NMR spectroscopy on protons, the signal intensity depends on the population difference between the two energy levels. The system is irradiated with a frequency, whose energy is equal to the difference in energy between these two energy levels. The transitions will be induced from the lower energy level to the higher and also in the reverse direction. Upward transitions absorb energy and downward transitions release energy. The probability for transitions in either direction is the same. The number of transitions in either direction is determined by multiplying the initial level population by a probability. A population of the initial level is determined by thermodynamics in accordance with the *Boltzmann* distribution.

The nuclei, which are in a lower energy state, can absorb radiation. This is a small proportion of nuclei. At absorption of radiation, a nucleus jumps to the higher energy state. When the populations of the energy levels are the same, the number of transitions in either direction will also be the same. Therefore, the absorption and release of energy will balance each other to zero. Then, there will be no further absorption of radiation in this case. Thus, the spin system will be *saturated*. However, a possibility of saturation means that the *relaxation processes* would occur to return nuclei to the lower energy state.

In order to avoid saturation, the populations of the levels should go back to the states at equilibrium. It does mean that nuclei from the higher energy state return to the state with lower energy. Emission of radiation is insignificant because the probability of re-emission of photons varies with the cube of the frequency. At radio frequencies, re-emission is negligible. Commonly, the NMR researcher would like relaxation rates to be fast but not too fast. If the relaxation rate is fast, then saturation is reduced. If the relaxation rate is too fast, line-broadening in the resultant NMR spectrum is observed. There are two major relaxation processes: (1) *spin-lattice (longitudinal)* relaxation and (2) *spin-spin (transverse)* relaxation.

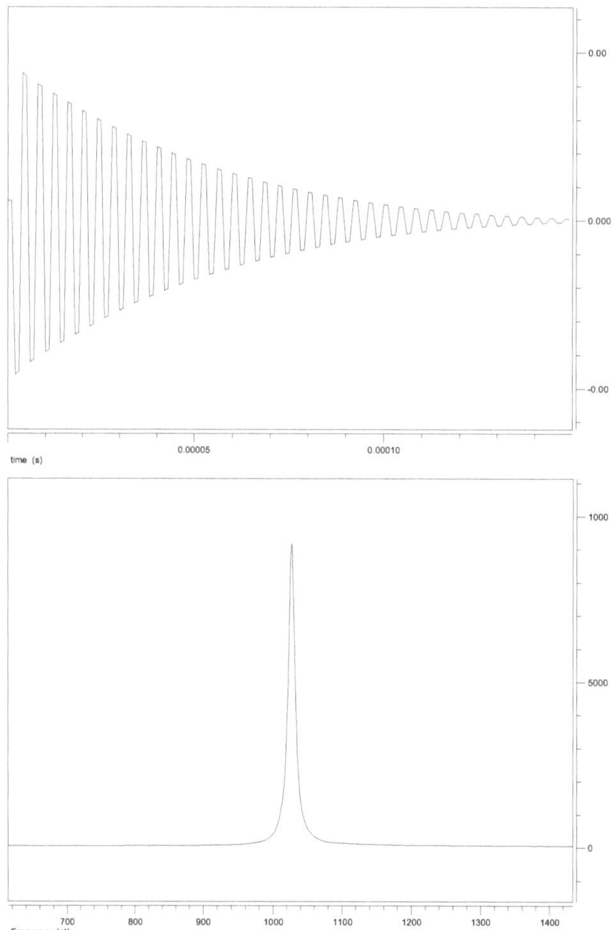

Fig. (1). FID signal (*top*) and ^{1}H NMR spectrum in frequency domain (*bottom*) for water protons of cement paste (water to cement ratio of 0.4). Age =3 days. T=298 K. Frequency for protons is 400 MHz. The Fourier transform between time domain (*top*) and frequency domain (*bottom*) has been performed using software *MestreC*. Frequency scale (*bottom*) is presented in points (pt): 50 pts correspond to 12 kHz.

Spin-Lattice (Longitudinal) Relaxation

All nuclei in the sample that are not observable are considered as *lattice*. Nuclei in the lattice are in vibrational and rotational motion which creates a complex magnetic field. The magnetic field caused by motion of nuclei within the lattice is called the *lattice field*. The components of the lattice field, which are equal in frequency and phase to the *Larmor* frequency of the considered nuclei, can interact with nuclei in the higher energy state, and cause them to lose energy and to return to the lower state. The effect of a resonant radio frequency (RF) pulse is to disturb the system of spins from its equilibrium state. The equilibrium is considered as a state of polarization with magnetization M_0 directed along the longitudinal magnetic field B_0. The restoration of the equilibrium is named longitudinal relaxation. This process can be described by eq. (3):

$$\frac{dM_z}{dt} = \frac{-(M_z - M_0)}{T_1} \tag{3}$$

The relaxation time constant, T_1, describes the average lifetime of nuclei in the higher energy state. T_1 is typically in the range of 0.1 - 20 sec for protons in non-viscous liquids and other dielectric materials at room temperatures. A larger T_1 indicates a slower or more inefficient spin relaxation [1 - 3]. T_1 is dependent on the *gyromagnetic ratio* of nucleus and the mobility of lattice. As mobility increases, the vibrational and rotational frequencies increase, making it more likely for a component of the lattice field to be able to interact with the excited nuclei. However, at extremely high mobility, the probability of a component of the lattice field being able to interact with the excited nuclei decreases. The efficiency of spin-lattice relaxation depends on factors that influence molecular movement in the lattice, such as viscosity and temperature. The relaxation process is kinetically first order. The applications of relaxation times are very important as they are used in the determination of the mobility of macromolecules and studying intra- and intermolecular interactions. Understanding the relaxation processes and relaxation times is essential in MRI studies. In order to monitor the different MR signals in various soft tissues, the pulse sequences, which include T_1-weighted sequences, are applied. T_1-weighted sequences are designed for obtaining the images and evaluation of anatomic structures.

The longitudinal relaxation times are often measured using the inversion-recovery pulse sequence (180°–τ–90°) [5, 8, 10 - 12]. Here τ is the time gap between 180° and 90° pulses. With the aid of the 1st *RF* pulse, the equilibrium magnetization M_0 rotates by 180° and becomes oriented along the $-Z'$-axis. During spin-lattice relaxation process, the magnetization changes from $-M_0$ *via* 0 towards the equilibrium magnetization M_0. After time τ, if a $90°_x$ pulse (with B_1 oriented along the X'-axis) is applied to the system, a magnetization will be oriented along the $-Y'(Y')$-axes and can be measured. When τ is varied, then experimental dependence $M_\tau = f(\tau)$ can be obtained and fitted by the law of M_τ =

M_0×[1–2×exp(–τ/T_1)] for single exponential spin-lattice relaxation. In the case of several components, $M_\tau = \sum M_{0i}$×[1–2×exp(–τ/T_{1i})] (where summation is doing for all relaxation components whereas *i* is the number of relaxation component). T_1 values are then calculated by performing nonlinear least squares fit to the data [3, 5, 8, 12]. Fig. (**2**) shows the result of one particular inversion recovery NMR experiment with fitting magnetization curve by two exponential components.

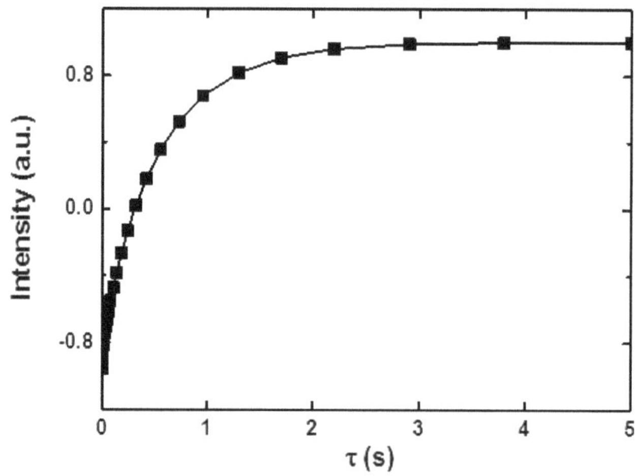

Fig. (2). The behaviour of longitudinal magnetization (normalized to M_0) *vs.* time registered in the inversion-recovery experiment (180°–*τ*–90°) on the *Bombyx mori* silk with water content of 0.18 g H_2O per g dry mass.T=298 K. Frequency is 400 MHz. Solid line is the fit of the experimental data set by the sum of two components (*a* and *b*): $M_\tau = M_{0a}$×[1–2×exp(–τ/T_{1a})] + M_{0b}×[1–2×exp(–τ/T_{1b})]. Spin-lattice relaxation time T_{1a} for slow relaxing component was 0.53 s, whereas fast relaxing component T_{1b} = 9 ms. Silk fibers from the *Bombyx mori* silkworm have two protein-monofilaments (brins) embedded in a glue-like sericin coating [13, 14]. The brins are fibroin filaments made up of bundles of nanofibrils, circa 5 nm in diameter, with a bundle diameter of around 100 nm. The nanofibrils are oriented parallel to the axis of the fiber, and are interacting strongly with each other [14].

Spin-Spin (Transverse) Relaxation

Spin-spin relaxation describes the interaction between neighbouring nuclei with identical precessional frequencies but differing magnetic quantum states. In this situation, the nuclei can exchange quantum states. A nucleus in the lower energy level will be excited, while the excited nucleus relaxes to the lower energy state. There is no net change in the populations of the energy states, but the average lifetime of a nucleus in the excited state will decrease. This can result in line-broadening. More details can be found in literature [1 - 5].

Transverse (spin-spin) relaxation is characterized by time constant T_2. Transverse relaxation is a process whereby nuclear spins come to thermal equilibrium between themselves. At room temperatures, the T_2 values for different materials are usually in the range of 10 μs - 10 sec. It is always true for $T_2 \leq T_1$.

In solutions $T_2 \approx T_1$, and in solids $T_2 << T_1$.

The values of T_2 are much less dependent on field strength, than T_1 values. Transverse magnetization corresponds to a state of phase coherence between the nuclear spin states. Transverse relaxation is unlike the longitudinal relaxation. It is sensitive to the interaction which results in dephasing nuclear spins. A description of spin-spin relaxation is presented in eq. (4):

$$\frac{dM_{x,y}}{dt} = \frac{-M_{x,y}}{T_2} \quad \textbf{(4)}$$

For measurement of T_2, it is possible to use *Han* pulse sequence (90°–τ–180°). This is a sequence which creates a signal of spin-echo at the moment $t = 2\tau$ [3, 5]. When interval 2τ is varied, an experimental dependence $M_{2\tau} = f(2\tau)$ can be obtained and modelled to calculate the T_2 [3]. The modification of *Han* method is *Carr-Purcell* (CP) pulse sequence (90°–τ_{cp}–180°–2τ_{cp}–180°–2τ_{cp}– 180°–...) which contains π pulses train. The *Carr-Purcell-Meiboom-Gill* (CPMG) sequence is the next modification of CP method that uses phase shift of π pulses (by 90°) with respect to initial 90° pulse [5, 8, 11, 12]. Fig. (**3**) shows an example of spin-echo intensities obtained in one particular CPMG experiment on wood sample.

Normally, in these CPMG NMR studies, it is necessary to model the decays of spin-echo intensities by a sum of several relaxation components [15]. Alternatively, a distribution of T_2 relaxation times, which could reconstruct spin-echo decay very closed to that from CPMG experiment, should be found. Conversion of the relaxation signal into a continuous distribution of T_2 relaxation components is based on inverse Laplace transform (ILT) [8, 15 - 18], *i.e.,* the inverse Laplace transform is used for extraction of $f(T_2)$ [19]. Thus, the data can be fitted applying ILT in one direction. The probability density $f(T_2)$ is calculated as follows from the eq. (5) for the spin-echo signal M_t presented according to [8, 16, 19]:

$$M_t = \int_0^{+\infty} e^{-t/T_2} f(T_2) dT_2 \quad \textbf{(5)}$$

The eq. (5) has the form of a Laplace transform. Fig. (**4**) presents an example of ILT application on data set obtained in CPMG experiment on wood.

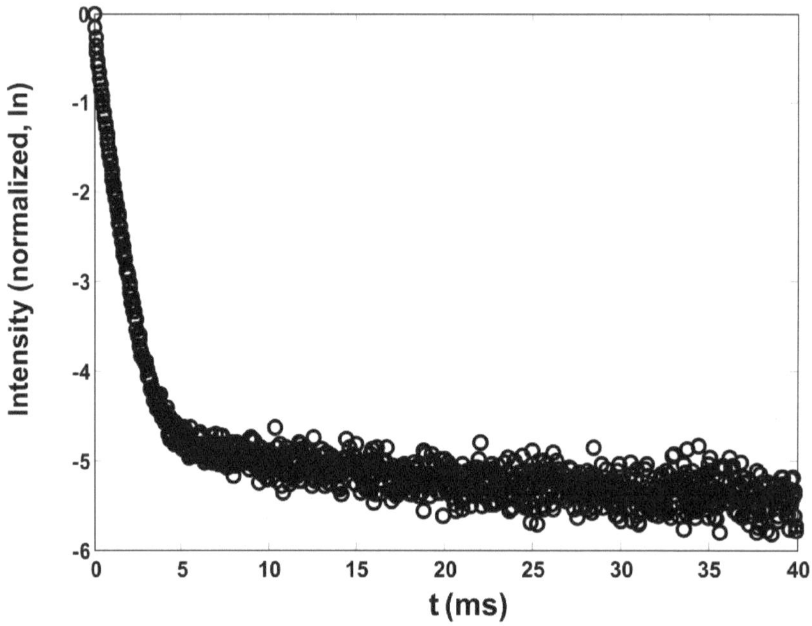

Fig. (3). The behaviour of spin-echo signal with time measured in CPMG experiment on the piece of wood with hydration level (HL) of 0.03 g H_2O per g dry mass. Wood samples were prepared from fresh pieces of felled mature spruce. The samples were cut from pieces of wood as follows: they were roughly right cube size (6×6×6 mm^3) that allowed to rotate a wood piece by 90° in NMR tube (10 mm in diameter) leaving the direction of gradient to main magnetic field as constant. T=298 K. Frequency is 400 MHz. CPMG curve could be fitted by the sum of three exponential components. The slowest T_{2a} component (though with very low intensity) was about 60-70 ms. The two fast relaxing components in the decay of spin-echo signal had the spin-spin relaxation times of T_{2b} =1-2 ms and T_{2c}= 0.15-0.20 ms.

EXPERIMENTAL MATERIALS AND METHODS

The following sections give preliminary information about those NMR methods that applied in studying T_1, T_2, cross-relaxation, DQF NMR, and NMR diffusion in natural silk *Bombyx mori*, collagens, wood, and polymer materials described in other chapters. It could be useful at reading any particular chapter in studying porous materials to return to this section to refresh these details. For more detail characterization of the materials, other physical methods were additionally applied [11, 15, 23 - 27]. Fig. (**5**) presents an example of electron microscope images of the studied *Bombyx mori* silk fibers, typically 15-30 μm in diameter.

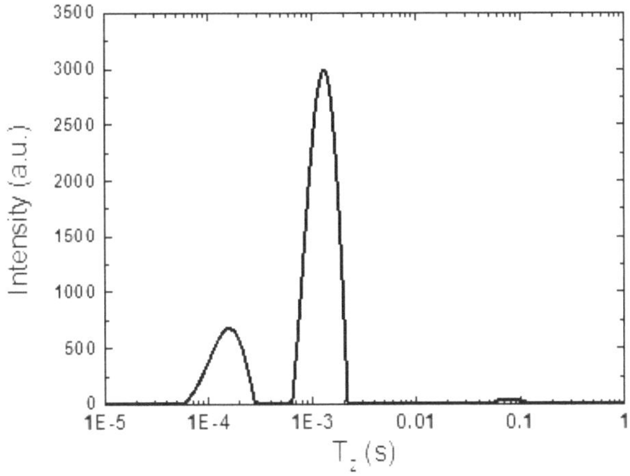

Fig. (4). T_2 distribution obtained in ILT application on the data set measured in ^{1}H CPMG NMR experiment on the piece of wood with HL = 0.03 g H_2O per g dry mass. T=298 K. Frequency is 400 MHz. The peak with T_2 ≈70-80 ms characterizing mobile water is practically absent. The sample have been dried to get the necessary HL=0.03 g/g which was lower than fiber saturation point (FSP) [20 - 22]. The FSP is the point at which free water has been removed from the wood and there is only bound water left [20, 21]. When HL is above the FSP, in addition to the fast relaxing T_2 components, a slowly relaxing component with a noticeable intensity (with T_2 in the range of 50-120 ms) is also registered [15].

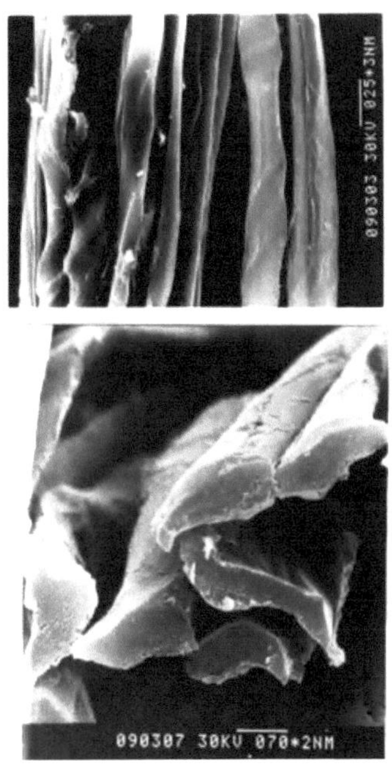

Fig. (5). Scanning electron microscope images of the surface of *Bombyx mori* natural silk fibers and their cross sections. The images were taken with a Hitachi S-450 scanning electron microscope (Japan) at an electron accelerating voltage of 30 kV using published procedures of sample fixation and preparation for studies [26, 27].

NMR Techniques for Studying T_1, T_2, and Cross-Relaxation

In the single pulse ^{1}H NMR experiments at 400 MHz, the FID signal after 90° pulse (length of the pulse = 5-6 μs) in hydrated *Bombyx mori* silk was converted from the time domain to the frequency-domain by FT and was recorded as a single line of proton resonance signal. In these experiments on hydrated silk fibers, the investigation of water was in the area of main interest [15]. Therefore, experimental parameters were chosen with such an approach that only protons of water were measured in the study [10, 11]. The transverse (spin-spin) relaxation time T_2 in the *Bombyx mori* silk was measured using the CPMG pulse sequence [5, 8, 11]. The T_2 value was also additionally estimated by analyzing the shape of the resonance signal line using the relationship between the spin-spin relaxation time T_2 and the *Lorentzian* width of the NMR spectrum as $T_2 = 1/\pi\omega_L$ [5, 8, 11, 15, 23]. The spin-lattice relaxation time T_1 in the natural silk samples was measured by the inversion-recovery (IR) method using the pulse sequence (180°–τ–90°) at the pulse sequence repetition time $\geq 5T_1$ [5, 12, 15]. The period of $5T_1$ was waited between consecutive scans of pulse sequence in order to ensure that the fully recovered signal was acquired. When t is varied, then experimental dependence of magnetization $M_t = f(t)$ can be obtained and fitted by the law $M_t = \sum M_{0i} \times [1-2\times\exp(-t/T_{1i})]$, where i is the number of relaxation component.

For the *Bombyx mori* silk samples with HL = 0.18, the characteristic time T_1 curve was modelled by two exponential components T_1' = 530 ms and T_1" = 9 ms with relative populations of 92 and 8%, respectively (Fig. **2**). With an increase in the duration of the π pulse, the population of the long component decreased and the population of the short component increased [11, 12].

The NMR studies on natural silk samples were also performed with pulsed field NMR relaxometer with an operating frequency for protons of 20 MHz [15, 23]. The parameters in these experiments were chosen to record all points (to observe the protons of both water and biopolymer). The FID signal of silk samples was found to comprise several distinguishable components [23]. In the first approximation, this signal can be regarded as consisting of a rapidly relaxing component (which characterized protons of highly limited mobility) and slowly decaying component (which represented more mobile protons). The results obtained on silk at 20 MHz were considered as both FID in time domain and spectrum after FID conversion to frequency domain. Thus, both these approaches characterized mobile protons and protons with limited mobility [5, 15].

For the collagen samples of lesser spotted dogfish *Scyliorhinus canicula*, NMR relaxation (T_1 and T_2) was studied at a proton resonance frequency of 20 MHz [15, 24] and at HL $\geq$ 0.6 g H_2O per g dry mass. The results obtained provided additional information on relaxation mechanism and the properties of bound water in biomaterials and were analysed in terms of known models of the water distribution in natural biopolymers [12, 15, 27]. The FID signal of *Scyliorhinus canicula* collagen samples consisted of fast and slow components similar to those detected in hydrated natural silk [11, 12, 23]. The slowly relaxing component characterised the hydration water with mobile and exchangeable groups of macromolecules whereas the fast relaxing component described low mobility protons of macromolecular collagen chains. The ratio of the amplitude of the slowly relaxing component to the amplitude of the fast relaxing component ranged from 1 to 2.5 and correlated with the water content of the studied samples similarly to the data of published works [15, 24 - 27]. The spin-spin relaxation time T_2 in lesser spotted dogfish collagen samples was measured using the CPMG pulse sequence at a time of $2\tau_{cp}$ = 100 μs between π pulses [24]. The spin-lattice relaxation time was measured by IR method (180°–τ–90°) [5, 8, 12, 15, 28]. In all the NMR experiments, care was taken to see that the sample with measurable stuff was contained within the volume of the probe coil to ensure good radiofrequency field homogeneity.

The ^{1}H NMR experiments on C15m and C8y collagens (these are collagen fibers of the aponeurosis of 15 month old calf and 8 years old steer, respectively [15, 17, 28]) were performed on a 400 MHz spectrometer (Bruker) at room temperature. These collagens were chosen according to different numbers of nonreducible covalent cross-links, which increase during the life of the animal. The extracted tissues were dried in air after washing with acetone for 4 times to remove fat. Additionally, the tissue samples were drying at 106 °C for 24 h. Collagen content (hydroxyproline concentration) was of 65% dry matter as in the published works [15, 17, 45]. Normally, three different samples were studied for each type of collagen, and the results obtained were averaged. These measurements were carried out at experimental conditions when transverse magnetization of collagen protons decayed very fast to zero (whereas the transverse water magnetization merely partially decayed), and only protons of water were measured in the study [17, 28]. This approach was usually used when only the investigation of water in hydrated collagen fibers was of main interest [15, 17, 26 - 28, 45]. The longitudinal relaxation times in collagens were measured using IR pulse sequence (180°–τ–90°) [5, 11, 17]. The T_1 values were calculated by performing the nonlinear least squares fit to the data. This T_1 experiment observed two exponentials. The measurements of T_2 have been carried out using CPMG pulse sequence with varying values of $2\tau_{cp}$ in π pulses train [15, 17]. CPMG curves were fitted as biexponential curves using in-house *MatLab*® codes with nonlinear least squared algorithm. The *Freeman–Hill* modification of the *Carr–Purcell* sequence (*Carr–Purcell–Freeman–Hill*, or CPFH) has been used also in some T_2 experiments on collagens with a pulse gap interval from 50 μs to 1 ms [15, 29]. It employed a (0°, 180°) alternation of the π pulse phases. The CPFH experiments discovered two exponentials.

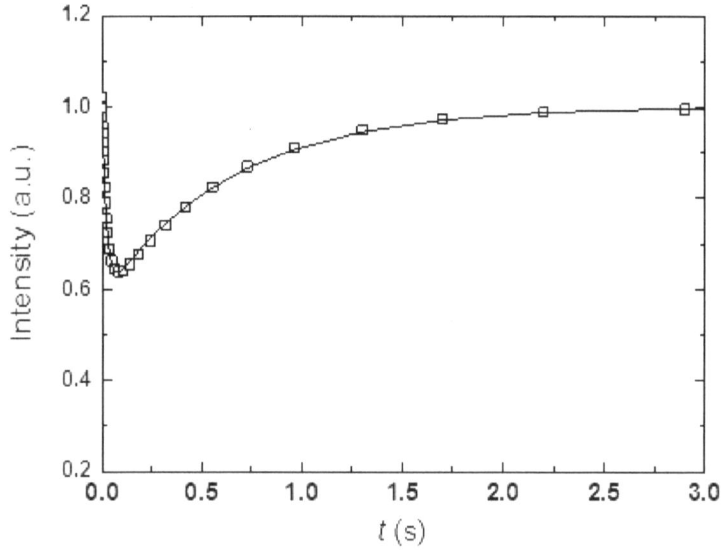

Fig. (6). Dependence of ^{1}H signal of water (open squares) on the spacing *t* between the 2nd and 3rd 90° pulses in *Goldman-Shen* pulse sequence ($90°_x–\tau_o–90°_{-x}–t–90°_x$) in cross-relaxation experiment for randomly oriented collagen fibers of C8y at HL = 0.6 g H_2O per g dry mass [28]. Solid line is the fit of the equations (eq. (7) to eq. (12)) to the experimental data according to the theory [12, 15, 17, 28, 30]. The dip on the curve shows the existence of a cross relaxation. $T = 298$ K. The frequency is 400 MHz.

The studies on collagens showed that there is a proton exchange between the water and exchangeable protons of macromolecular pool. The changes in the intensity of slow relaxing component confirmed that [17, 28]. This cross-relaxation effect is observed in many macromolecular systems with low water content [11, 12, 23, 28, 30]. Cross-relaxation experiments for collagens were implemented according to the *Goldman-Shen* (GS) pulse sequence ($90°_x–\tau_o–90°_{-x}–t–90°_x$) used earlier in the works [12, 15, 17, 28]. Fig. (**6**) shows a typical data set obtained from the GS experiment on collagen tissue. The first two pulses in GS experiment created the different conditions for longitudinal magnetizations of water protons and collagen protons. After the first time gap of $\tau_o = 65$ μs, the transverse magnetization of collagen decays to zero whereas the transverse magnetization of water doesn't decay. The second pulse (with opposite phase to the first one) rotates water proton magnetization back to the steady magnetic field. In the case of effect of cross-relaxation, longitudinal collagen and water magnetization will change during the time *t*. After the third $90°_x$ pulse, a measurement of free induction decay as a collection of transverse magnetization of water has been done. The implementation of this pulse sequence was done in the studies of hydrated silk *Bombyx mori*, collagen fibers of different ages and other materials [12, 15, 17, 28, 30, 31].

Pulsed Field Gradient (PFG) NMR in One- and Two-Dimensional Studies

Pulsed field gradient (PFG) NMR studies have been realized in one- and two-dimensions. In one-dimensional case, the three pulse stimulated-echo (STE) sequence [6, 8, 10, 15, 17, 32] (Fig. 7, *top*) was used with storage times τ_2 and encoding times $\Delta = \tau_1 + \tau_2$. In this pulse sequence, τ_1 is the gap between the 1st and 2nd, and τ_2 is the interval between the 2nd and 3rd 90° RF pulses. Gradient pulses of duration $\delta = 1$ ms and maximum amplitude $G_{max} = 80$ G/cm have been applied. The 90° RF excitation pulse was established as 10 μs. In order to set appropriate repetition time in the diffusion experiment, the T_1 values of the water signal from each sample were estimated by IR NMR technique [5, 15, 28].

When the NMR PFG technique is applied to heterogeneous systems, it is often necessary to take into account the existence of internal magnetic field gradients because of the heterogeneity in magnetic susceptibility of the sample [33 - 35]. The effects of internal magnetic field (susceptibility) gradients can be negated, for example, by the *Cotts* sequences [33]. *Cotts* introduced bipolar gradient pairs separated by additional 180° pulses instead each gradient pulse of the STE sequence. The '13-intervals sequence' of *Cotts* [33] using a bipolar pair of 1 ms gradient pulses in place of each 2 ms pulse has been implemented in collagen studies [17] to compare the results of STE and *Cotts* pulse sequence on the same samples.

Two-dimensional diffusion-diffusion correlation experiments (DDCOSY) [8, 15, 17, 35 - 37] comprising of two spin-echo pulse sequences with two pairs of PFG were carried out with encoding times of 6 ms and gradients in the *X*-directions (Fig. 7, *bottom*) or in both *X*- and *Y*-directions. Echo decays were inverted using the inverse Laplace transform (ILT) algorithm according to Venkataramanan *et al.* [38].

If the NMR echo intensity is recorded in presence of a pair of field gradient pulses of strength *G*, duration δ, and separation Δ, then it is smaller than the echo intensity in the absence of the gradients because of diffusive attenuation. According to the published works [6, 8, 15, 39 - 41], the amplitude of echo depends on the gradient factor $b = q^2(\Delta - \delta/3)$ and the echo time $(\tau_2 + 2\tau_1)$ as: $I(\tau_1,\tau_2,b)/I(0) = \exp(-\tau_2/T_1)\times\exp(-2\tau_1/T_2)\times\exp(-bD)$. So, for molecules diffusing in liquids, the additional echo-attenuation $I(G)/I^*(0)$ is given in eq.(6), where $q = \gamma G\delta$ and γ is the (^{1}H) nuclear *gyromagnetic ratio.*

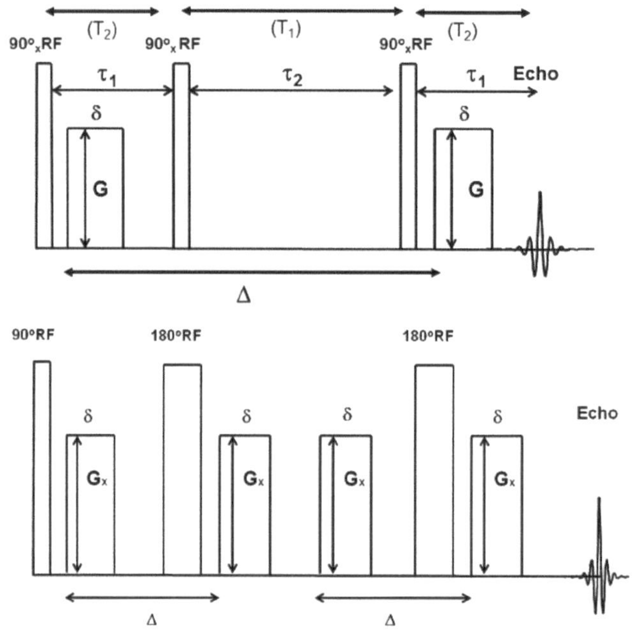

Fig. (7). Stimulated-echo (STE) pulse sequence for one-dimensional PFG experiments *(top)*; 2D spin-echo (DDCOSY) pulse sequence with two pairs of collinear gradient pulses G_x (length δ and diffusion time Δ) (*bottom*).

$$\frac{I(G)}{I^*(0)} = \exp[-q^2(\Delta - \frac{\delta}{3})D] \tag{6}$$

The eq.(6) has also been applied to systems undergoing restricted diffusion [8, 39 - 42]. Then, the diffusion coefficient D in eq. (6) should be replaced by an effective or apparent diffusion coefficient D_{app}. Sometimes the total water signal in diffusion studies contains a small water fraction bound to macromolecules by strong hydrogen bonding. The diffusion of this bound water is very slow relative to the diffusion time Δ. Therefore, one considers the signal of this bound water as a constant in NMR diffusion experiment [42].

Cross-Relaxation in PFG NMR Studies

In STE diffusion experiment, longitudinal relaxation has a place during diffusion interval between the 2nd and 3rd RF pulses (Fig. **7**, *top*). Cross-relaxation is driven by the difference between the deviations from equilibrium in both phases. According to Ref. [30], the cross-relaxation rate is depending on q. Therefore, the rate of cross-relaxation will depend on gradient *G*. The equations, which include exchange terms in *Bloch* equations for *Z*-magnetization of both phases, are presented anywhere [12, 28, 30]. The authors of publication [30] extended the *Bloch* equations for *Z*-magnetization in order to describe the influence of cross-relaxation on the attenuation of the STE signal by diffusion. Only longitudinal magnetization is subject to the cross-relaxation. Therefore, only the behaviour during interval τ_2 should be described, *i.e.*, when longitudinal relaxation has a place (Fig. **7**, *top*) [12, 30].

The data were analysed in terms of the two-site model [11, 12, 17, 28, 43]: (Fig. **8**). R_{1w} and R_{1c} are the intrinsic longitudinal relaxation rates, k_w and k_c are the cross-relaxation rates, p_w and p_c are the proton fractions ($p_w + p_c = 1$) of water and collagen proton pools, respectively. The dipolar cross-relaxation rate is normally measured by implementing GS experiment [11, 12, 15, 17, 25, 30, 43, 44] resulting in data set as it is showed in Fig. (**6**). If $M_w(t)$ is considered as the longitudinal magnetization of water pool, then the time dependent normalized deviation $\Delta m_w = (M_w(t) - M_w^{eq})/M_w^{eq}$ from the equilibrium value (M_w^{eq}) of longitudinal magnetization of the water pool can be presented [12, 17, 30] as shown in the following series equations (7) - (12):

$$\Delta m_w = c^+ e^{-R^+ t} + c^- e^{-R^- t} \tag{7}$$

$$2R^+ = k_w + R_{1w} + k_c + R_{1c} + \left[\left(k_w + R_{1w} - k_c - R_{1c} \right)^2 + 4k_w k_c \right]^{1/2} \tag{8}$$

$$2R^- = k_w + R_{1w} + k_c + R_{1c} - \left[\left(k_w + R_{1w} - k_c - R_{1c} \right)^2 + 4k_w k_c \right]^{1/2} \tag{9}$$

$$c^+ = +\Delta m_w(\tau_0) \frac{k_w + R_{1w} - R^-}{R^+ - R^-} - \Delta m_c(\tau_0) \frac{k_w}{R^+ - R^-} \tag{10}$$

$$c^- = -\Delta m_w(\tau_0) \frac{k_w + R_{1w} - R^+}{R^+ - R^-} + \Delta m_c(\tau_0) \frac{k_w}{R^+ - R^-} \tag{11}$$

$$\Delta m_w(\tau_0) = e^{-\pi\omega_L \tau_0 - \left(\frac{\pi\omega_G \tau_0}{2(\ln 2)^{1/2}} \right)^2} - 1 \tag{12}$$

The water NMR signal is neither purely exponential nor purely *Gaussian.* A line shape of the signal was described by *a Voigt* function which is produced by a *Gaussian* broadening of the *Lorentzian* peak [12, 17, 28]. ω_L and ω_G are the *Lorentzian* and *Gaussian* full widths at half-maximum height of the peaks in the frequency domain, respectively. It was experimentally found that at $\tau_0 = 65$ μs the transverse magnetization of collagen $M_c(\tau_0)\approx 0$ (in comparison with $M_w(\tau_0)$) and the normalized deviation $\Delta m_c(\tau_0) = [M_c(\tau_0) - M_c^{eq}]/M_c^{eq}$ from the equilibrium value of longitudinal magnetization of the collagen pool, is equal to –1 [12, 28]. Fitting the equations (7)-(12) to the GS experiment data (Fig. **6**) yields estimates of k_w, R_{1w}, k_c and R_{1c}. The authors of the work [30] evaluated the effect of cross-relaxation on the stimulated echo amplitude when one measures self-diffusion coefficient values that decrease with increasing diffusion times. Their complete expression for stimulated-echo-attenuation in the presence of cross-relaxation was derived as solution of differential *Bloch* equations for *Z*-magnetization of both phases (water and macromolecule) including exchange terms. Final expression displayed in the work [30] is rather unwieldy and is not given here. It contains all the parameters of STE PFG experiment, including the diffusion constant, and a combination of four cross-relaxation parameters.

When the parameters k_w, R_{1w}, k_c and R_{1c} are known, the diffusion coefficient can be determined by a nonlinear regression analysis fitting the echo-attenuation curve to derived expression [15, 30]. The diffusion coefficients (D_{app}) in C15m and C8y collagen tissues have been measured for two perpendicular directions (*X, Z*) applying the gradient of magnetic field with 0° or 90° to the static magnetic field B_0 and taking into account the cross-relaxation factor at analysis of the data obtained [15, 17].

Double-Quantum-Filtered (DQF) NMR Spectroscopy

Water bound to fibrous proteins (collagen or other natural biomaterials with low HL value) is an example of anisotropic motion. The fibers, porous systems, and other materials with anisotropic motion of molecules could be studied by the multiple-quantum-filtered (MQF) NMR techniques [15, 25, 45 - 52]. It has been demonstrated that filtering the NMR signal with the aid of these techniques enables the observation of the contributions to the FID that originate exclusively from anisotropic media [25, 45 - 47]. MQF NMR is applied for investigating the systems, in which the residual dipolar interaction may be of local or macroscopic nature [46 - 50].

It has been demonstrated that ^{1}H double-quantum-filtered (DQF) NMR spectroscopy is a sensitive method for depicting biological tissues associated with the ordered structure [45 - 47]. Because dipolar interaction completely averages to zero in isotropic media, a DQF NMR signal is obtained only in the presence of an ordered structure. DQF spectra of water molecules have been detected in collagen containing tissues (ligaments, tendons, cartilage, skin, blood vessels) and natural silk *B. mori* [15, 25, 45 - 51]. The DQ coherences in the collagen based tissues are formed as a result of interaction with the collagen fibers. According to the published data [25, 47 - 50], a dependence of the DQF signal on the creation time and the value of creation time, at which a maximal DQF signal is registered, can be considered as a characteristic of each tissue. Therefore, it would be reasonable to compare the signals from collagen fibers with different cross-links, *i.e.*, studying connective tissues of different ages and performing the measurements at different creation times.

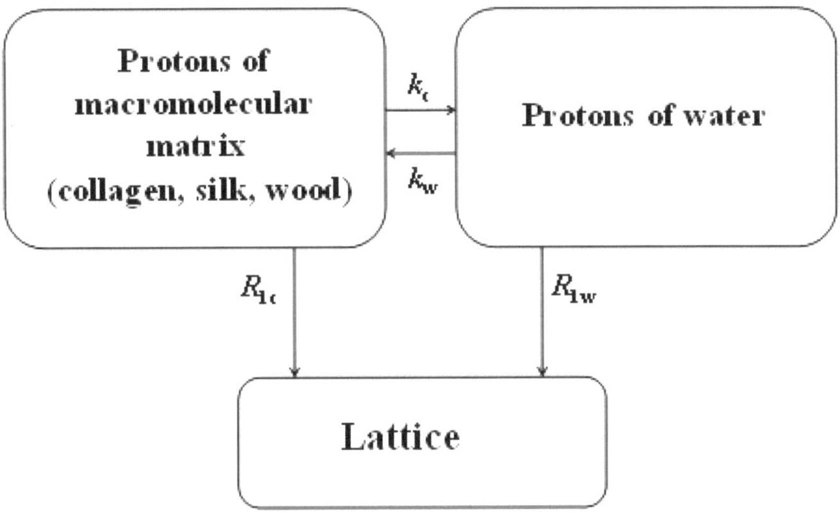

Fig. (8). Two phase model [12, 15, 30] that was used for estimation of cross-relaxation effects. R_{1w} and R_{1c} are the rates of longitudinal (spin-lattice) relaxation in each phase (*e.g.*, water protons and collagen protons, respectively). k_w and k_c are the cross-relaxation rates (they characterize a rate of magnetisation transfer from one proton phase to another one), $p_w=k_c/(k_w+k_c)$, $p_c=k_w/(k_w+k_c)$ are the populations of water and collagen, respectively. $p_w+p_c=1$. Water phase with population of p_w can include exchangeable polymer protons [28, 31].

DQF spectra of water molecules were detected in both young and adult collagen oriented fibers [15, 45] indicating the residual proton-proton dipolar interaction and anisotropy of water molecules. The dependence of the DQF signal on the creation time (τ_{creat}) had a maximal signal at $(\tau_{creat})_{max}$. The detection of DQF NMR signal was done using the following pulse sequence: $90°_{\varphi}$–$(\tau_{creat}/2)$–$180°_{\varphi 1}$–$(\tau_{creat}/2)$ –$90°_{\varphi}$–t_1–$90°_{\varphi 2}$–t_2 (acquisition) [15, 25, 46, 48] (Fig. **9**).

As a result of the residual dipolar interaction, double rank spherical tensor operators ($T_{2,+1}$, $T_{2,-1}$) are created during the creation time τ_{creat} [15, 48, 51, 52]. The

180° pulse refocuses frequency shifts and magnetic field inhomogeneities. To describe the processes during this pulse sequence, one used irreducible spherical tensor operator basis $T_{l,p}$ (l, a rank of the tensor; p, a coherence of the tensor with $-l \leq p \leq l$) [46]. Residual dipolar interactions (RDIs) result in the formation of tensors of high rank ($T_{l,\pm1}$, l =2, ...,) which can be detected by MQF NMR methods [15, 25, 47 - 52]. These tensors are presented as combination of known spin operators I_z and $I_\pm = I_x \pm iI_y$ [15, 45, 46]. For instance, for two ½ spins, spherical tensor $T_{1,0}$ is presented as $I_z^a + I_z^b$, tensors $T_{1,\pm1}$ as $-(\pm2^{-1/2})\cdot(I_\pm^a + I_\pm^b)$, tensors $T_{2,\pm1}$ as $-(\pm½)\cdot(I_z^aI_\pm^b + I_\pm^aI_z^b)$, and tensors $T_{2,\pm2}$ as $½\cdot(I_\pm^aI_\pm^b)$.

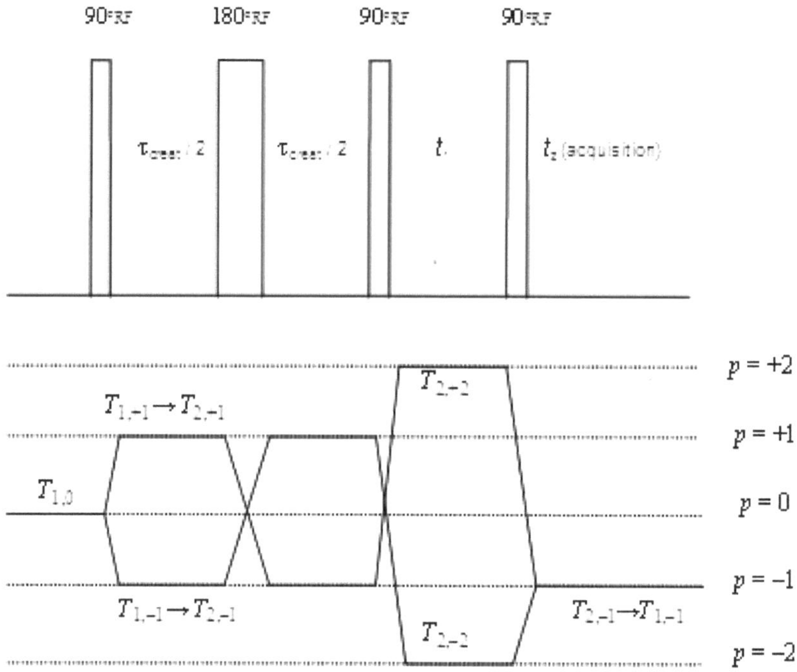

Fig. (9). DQF pulse sequence and the coherence-transfer pathway diagram. $T_{l,p}$ are spherical tensors of rank l and coherence p. Each pathway is chosen by the corresponding phase cycle [15, 25, 45 - 47].

The second 90° pulse transforms the second rank tensors into double quantum coherence tensors ($T_{2,+2}$, $T_{2,-2}$) that are selected using cycling of the phase, φ. The third 90° pulse transforms the coherence back to −1. Therefore, it may be detected during acquisition. The detected DQ signal of water results from anisotropic molecular motion. The DQF NMR signals were measured by varying the DQ evolution time, t_1 (from 10 μs to 5 ms). The intensity of DQF NMR signal decreased with increase of evolution time (at fixed τ_{creat}) [15, 45, 53]. The DQF evolution time t_1 was kept as short as possible to avoid lost of signal due to relaxation in experiments with changing creation time. In other DQF NMR experiments, value τ_{creat} was varied at constant value of t_1 = 100 μs [45, 48, 51]. The line broadening constant (LB) was 10 Hz.

Typical DQF experiment was carried out with pulse sequence mentioned above and had phase cycle according to published data [45 - 47, 49, 51, 53, 54]. The resulting DQF signal for detectable transverse magnetization is given by [53, 55]: $M(\tau_{creat},t_1,t_2) = 0.15\times[\exp(-s_2\tau_{creat})-\exp(-s_1\tau_{creat})]\times\exp(-s_{dq}t_1)\times[\exp(-s_2t_2)-\exp(-s_1t_2)]$, where s_{dq} is double-quantum transverse relaxation rate, and s_1, s_2 are two single-quantum transverse relaxation rates. At constant values of t_1 and t_2, the detectable transverse magnetization signal was dependent only on relaxation rate constants s_1 and s_2 [51, 53 - 55] as $M_{xy} = A\times[\exp(-s_2\tau_{creat})-\exp(-s_1\tau_{creat})]$ at A = const. The time dependent free precession signal in DQF experiments is the difference between two exponentials rather than the weighted sum obtained by single pulse acquisition. The corresponding absorption spectrum is the difference between two *Lorentzians* with different widths [11, 46, 57], *i.e*, two *Lorentzians* are in *anti*-phase to each other. The relative line widths of the broad and narrow components depend on the ratio of the two relaxation rates. For example, the broad component may not be detected if the contribution of narrow resonance dominates ($s_1 >> s_2$) [52, 54, 55]. In the case of $s_1 = s_2$, the double-quantum filtered spectrum vanishes. Results from experiments at series of τ_{creat} values therefore enable calculation of the relaxation rate constants s_1 and s_2 [45, 53 - 57] using nonlinear least squares fit of the dependence of amplitude of the DQF signals on the creation time. The τ_{creat} values, which result in maximal intensity of DQF сигналов in the dependences of intensity $I_{DQF} = f(\tau_{creat})$, were calculated according to: $\tau_{max}=\ln(s_1/s_2)/(s_1-s_2)$ [15, 45, 53].

CONCLUDING REMARKS

Several NMR experimental techniques have been introduced describing briefly NMR basics and some applications. NMR diffusion studies have been considered taking into account the details of proton exchange and cross-relaxation in hydrated porous materials. A realisation of PFG NMR techniques in one- and two-dimensions has been considered. To study hydrated materials with anisotropic motion of water, the DQF NMR method has been presented. The details of experiments on some particular materials and simulation studies are considered in following chapters. Simulations of spectral (1D and 2D), GS-, DQF- and echo-attenuation curves were performed using home-made codes in *MatLab* software package.

REFERENCES

[1] A. Abragam, *The Principles of Nuclear Magnetism.* Clarendon Press: Oxford, 1961.

[2] B. Blümich, *Essential NMR (for Scientists and Engineers).* Springer, 2005.

[3] T.C. Farrar, and E.D. Becker, *Pulse and Fourier Transform NMR: Introduction to Theory and Methods*. Academic Press, 1971.

[4] J. Keeler, *Understanding NMR spectroscopy*. 2nd ed. John Wiley & Sons Ltd, 2010.

[5] V.V. Rodin, *Magnetic Resonance Methods*. Press MIPT: Moscow, 2004.

[6] P.T. Callaghan, *Principles of Nuclear Magnetic Resonance Microscopy*. Clarendon Press: Oxford, 2001.

[7] K.M. Sanders, and B.K. Hunter, *Modern NMR Spectroscopy*. 2nd ed. Oxford University Press: Oxford, 1993.

[8] P.T. Callaghan, *Translational Dynamics and Magnetic Resonance*. Oxford University Press: Oxford, 2011.
[http://dx.doi.org/10.1093/acprof:oso/9780199556984.001.0001]

[9] H. Friebolin, *Basic One- and Two-Dimensional NMR Spectroscopy* Wiley-VCH Verlag GmbH & Co. KGaA: Weinheim, 2011.

[10] V.V. Rodin, and D.P. Knight, "Water self-diffusion in natural silk fibers as determined by the pulsed magnetic field gradient method", *Biophysics*, vol. 48, no. 3, pp. 404-410, 2003. http://epubs.surrey.ac.uk/835871/
[PMID: 12815853]

[11] V.V. Rodin, G.M. Reznichenko, and E.L. Vasina, "Properties of natural polymer fibers with low water content", *Mater. Sci.*, no. 1, pp. 34-42, 2004.

[12] V.V. Rodin, and D.P. Knight, "Natural materials by NMR data: cross-relaxation in macromolecules of natural silk", *Mater. Sci.*, no. 10, pp. 16-21, 2003.

[13] G.J.G. Davies, D.P. Knight, and F. Vollrath, "Chitin in the silk gland ducts of the spider *Nephila edulis* and the silkworm *Bombyx mori*", *PLoS One*, vol. 8, no. 8, p. e73225, 2013.
[http://dx.doi.org/10.1371/journal.pone.0073225] [PMID: 24015298]

[14] O. Hakimi, D.P. Knight, F. Vollrath, and P. Vadgama, "Spider and mulberry silkworm silks as compatible biomaterials", *Compos., Part B Eng.*, vol. 38, no. 3, pp. 324-327, 2007.
[http://dx.doi.org/10.1016/j.compositesb.2006.06.012]

[15] V.V. Rodin, "Methods of Magnetic Resonance in Studying Natural Biomaterials", In: *Encyclopedia of Physical Organic Chemistry*, Zerong Wang, Ed., Chapter 53. vol. 4, part 4. John Wiley & Sons, Inc., 2017, pp. 2861-2908. Available from: https://www.wiley.com/en-us/Encyclopedia+of+Physical+Organic+Chemistry%2C+6+Volume+Set-p-9781118470459

[16] P. Berman, O. Levi, Y. Parmet, M. Saunders, and Z. Wiesman, "Laplace inversion of low-resolution NMR relaxometry data using sparse representation methods", *Concepts Magn. Reson., Part A*, vol. 42, no. 3, pp. 72-88, 2013.

[17] V.V. Rodin, and V.A. Nikerov, "NMR relaxation and PFG NMR studies of water dynamics in oriented collagen fibres with different degree of cross-linking", *Curr. Tissue Eng.*, vol. 3, no. 1, pp. 47-61, 2014.
[http://dx.doi.org/10.2174/2211542003666140626211652]

[18] P. Fantazzini, A. Maccotta, M. Gombia, C. Garavaglia, R.J.S. Brown, and M. Brai, "Solid–liquid nuclear magnetic resonance relaxation and signal amplitude relationships with ranking of seasoned softwoods and hardwoods", *J. Appl. Phys.*, vol. 100, pp. 0749071-0749077, 2006.

[19] H. Celik, M. Bouhrara, D.A. Reiter, K. W. Fishbein, and R.G. Spencer, "Stabilization of the inverse Laplace transform of multiexponential decay through introduction of a second dimension", *J. Magn. Reson.*, vol. 236, pp. 134-139, 2013.
[http://dx.doi.org/10.1016/j.jmr.2013.07.008]

[20] C.A.S. Hill, A.J. Norton, and G. Newman, "The water vapour sorption properties of Sitka Spruce

determined using a dynamic vapour sorption apparatus", *Wood Sci. Technol.,* vol. 44, pp. 497-514, 2010.
[http://dx.doi.org/10.1007/s00226-010-0305-y]

[21] V-V. Telkki, M. Yliniemi, and J. Jokisaari, "Moisture in softwoods: fiber saturation point, hydroxyl site content, and the amount of micropores as determined from NMR relaxation time distributions", *Holzforschung,* vol. 67, no. 3, pp. 291-300, 2013.
[http://dx.doi.org/10.1515/hf-2012-0057]

[22] J. Cox, P.J. McDonald, and B.A. Gardiner, "A study of water exchange in wood by means of 2D NMR relaxation correlation and exchange", *Holzforschung,* vol. 64, no. 2, pp. 259-266, 2010.
[http://dx.doi.org/10.1515/hf.2010.036]

[23] V.V. Rodin, G.M. Reznichenko, and E.L. Vasina, "Properties of natural silk fibers: deformation studies and NMR data", *Biophysics,* vol. 49, no. 6, pp. 918-926, 2004.
[PMID: 15612542]

[24] V.V. Rodin, "NMR Relaxation in investigation of water in hydrated collagen of Lesser Spotted Dogfish", *Biophysics,* vol. 50, no. 2, pp. 219-225, 2005.

[25] V.V. Rodin, and D.P. Knight, "Molecular mobility in natural polymers: *Bombyx mori* silk with low water content as studied by ^{1}H DQF NMR", *Biophysics,* vol. 49, no. 5, pp. 730-737, 2004.
[PMID: 15526463]

[26] V.V. Rodin, B.V. Sakharov, V.N. Gerasimov, and D.P. Knight, "NMR Relaxation in natural polymers (collagen spicules)", *Int. Polym. Sci. Technol,* vol. 28, no. 3, pp. 70-78, 2001.

[27] V.V. Rodin, B.V. Sakharov, V.N. Izmailova, and D.P. Knight, "Study of the collagen spicule structure by NMR relaxation and electron microscopy", *Russ. J. Biotechnol.,* no. 4, pp. 257-265, 2001.

[28] V.V. Rodin, L. Foucat, and J.P. Renou, "Natural polymers according to NMR data: cross-relaxation in collagen macromolecules from two samples of connective tissues", *Biophysics,* vol. 49, no. 4, pp. 563-571, 2004.
[PMID: 15458243]

[29] R. Freeman, and H.D.W. Hill, "High-resolution study of NMR spin echoes: J-Spectra", *J. Chem. Phys.,* vol. 54, pp. 301-313, 1971.
[http://dx.doi.org/10.1063/1.1674608]

[30] L.J.C. Peschier, J.A. Bouwstra, J. de Bleyser, H.E. Junginger, and J.C. Leyte, "Cross-relaxation effects in pulsed-field-gradient stimulated-echo measurements on water in a macromolecular matrix", *J. Magn. Reson. B.,* vol. 110, no. 2, pp. 150-157, 1996.
[http://dx.doi.org/10.1006/jmrb.1996.0024]

[31] M-C. Vackier, B.P. Hills, and D.N. Rutledge, "An NMR relaxation study of the state of water in gelatin gels", *J. Magn. Reson.,* vol. 138, no. 1, pp. 36-42, 1999.
[http://dx.doi.org/10.1006/jmre.1999.1730] [PMID: 10329223]

[32] V.V. Rodin, and T. Cosgrove, "Nuclear magnetic resonance study of water-polymer interactions and self-diffusion of water in polymer films", *OALJ: Chem. Mater. Sci,* vol. 3, no. 10, pp. 1-17, 2016.
[http://dx.doi.org/10.4236/oalib.1103018]

[33] R.M. Cotts, M.J.R Hoch, T. Sun, and J.T. Markert, "Pulsed field gradient stimulated echo methods for improved NMR diffusion measurements in heterogeneous systems", *J. Magn. Reson.,* vol. 83, no. 2, pp. 252-266, 1989.

[34] G.H. Sorland, B. Hafskjold, and O. Herstad, "A stimulated-echo method for diffusion measurements in heterogeneous media using pulsed field gradients", *J. Magn. Reson.,* vol. 124, no. 1, pp. 172-176, 1997.
[http://dx.doi.org/10.1006/jmre.1996.1029]

[35] V.V. Rodin, P.J. McDonald, and S. Zamani, "A nuclear magnetic resonance pulsed field gradient study of self-diffusion of water in hydrated cement pastes", *Diffus. Fundam.,* vol. 18, no. 3, pp. 1-7,

2013.

[36] P.T. Callaghan, and I. Furó, "Diffusion-diffusion correlation and exchange as a signature for local order and dynamics", *J. Chem. Phys.,* vol. 120, no. 8, pp. 4032-4038, 2004. [http://dx.doi.org/10.1063/1.1642604] [PMID: 15268569]

[37] V.V. Rodin, P.J. McDonald, and M. Jones, "Two-dimensional distribution function of diffusion in wood obtained using 2D Laplace inversion", *Appl. Phys. Math.,* no. 6, pp. 3-7, 2014.

[38] L. Venkataramanan, Yi-Q. Song, and M.D. Hurlimann, "Solving Fredholm integrals of the first kind with tensor product structure in 2 and 2.5 Dimensions", *IEEE Trans. Signal Process.,* vol. 50, no. 5, pp. 1017-1026, 2002. [http://dx.doi.org/10.1109/78.995059]

[39] J.E. Tanner, and E. Stejskal, "Restricted self-diffusion of protons in colloidal systems by the pulsed-gradient, spin-echo method", *J. Chem. Phys.,* vol. 49, no. 4, pp. 1768-1777, 1968. [http://dx.doi.org/10.1063/1.1670306]

[40] J.E. Tanner, "Transient diffusion in system partitioned by permeable barriers. Application to NMR measurements with a pulsed field gradient", *J. Chem. Phys.,* vol. 69, no. 4, pp. 1748-1754, 1978. [http://dx.doi.org/10.1063/1.436751]

[41] T. Cosgrove, V.V. Rodin, M. Murray, and R. Buscall, "Self-diffusion in solutions of carboxylated acrylic polymers as studied by pulsed field gradient NMR. 2. Diffusion of macromolecules", *J. Polym. Res.,* vol. 14, no. 3, pp. 175-180, 2007. [http://dx.doi.org/10.1007/s10965-006-9088-0]

[42] W. Wycoff, S. Pickup, B. Cutter, W. Miller, and T. Wong, "The determination of the cell size in wood by nuclear magnetic resonance diffusion techniques", *Wood Fiber Sci.,* vol. 32, no. 1, pp. 72-80, 2000.

[43] S. Peto, P. Gillis, and V.P. Henri, "Structure and dynamics of water in tendon from NMR relaxation measurements", *Biophys. J.,* vol. 57, no. 1, pp. 71-84, 1990. [http://dx.doi.org/10.1016/S0006-3495(90)82508-X] [PMID: 2297563]

[44] V.V. Rodin, "NMR-approaches in studying dynamic molecules in porous materials", *J. Mater. Sci. Eng.,* vol. 3, no. 3, p. 245, 2014.

[45] V.V. Rodin, L. Foucat, and J.P. Renou, "The dipolar interactions and dynamics of water molecules in collagen fibers from two connective tissues of different ages by ^{1}H double-quantum-filtered NMR spectroscopy", *Mater. Sci.,* no. 5, pp. 2-10, 2004.

[46] G. Navon, H. Shinar, U. Eliav, and Y. Seo, "Multiquantum filters and order in tissues", *NMR Biomed.,* vol. 14, no. 2, pp. 112-132, 2001. [http://dx.doi.org/10.1002/nbm.687] [PMID: 11320537]

[47] U. Eliav, M. Komlosh, P.J. Basser, and G. Navon, "Characterization and mapping of dipolar interactions within macromolecules in tissues using a combination of DQF, MT and UTE MRI", *NMR Biomed.,* vol. 25, no. 10, pp. 1152-1159, 2012. [http://dx.doi.org/10.1002/nbm.2784] [PMID: 22362643]

[48] U. Eliav, and G. Navon, "A study of dipolar interactions and dynamic processes of water molecules in tendon by ^{1}H and ^{2}H homonuclear and heteronuclear multiple-quantum-filtered NMR spectroscopy", *J. Magn. Reson.,* vol. 137, no. 2, pp. 295-310, 1999. [http://dx.doi.org/10.1006/jmre.1998.1681] [PMID: 10089163]

[49] G. Navon, U. Eliav, D.E. Demco, and B. Blümich, "Study of order and dynamic processes in tendon by NMR and MRI", *J. Magn. Reson. Imaging,* vol. 25, no. 2, pp. 362-380, 2007. [http://dx.doi.org/10.1002/jmri.20856] [PMID: 17260401]

[50] V.V. Rodin, "Translational dynamics of molecules in synthetic and natural polymers as studied by NMR", *27th ECIS Conference* 2013, Sofia, Bulgaria.

[51] Y. Seo, K. Ikoma, H. Takamiya, Y. Kusaka, L. Tsoref, U. Eliav, H. Shinar, and G. Navon, " ^{1}H

double-quantum-filtered MR imaging as a new tool for assessment of healing of the ruptured Achilles tendon", *Magn. Reson. Med.,* vol. 42, no. 5, pp. 884-889, 1999. [http://dx.doi.org/10.1002/(SICI)1522-2594(199911)42:5<884::AID-MRM7>3.0.CO;2-0] [PMID: 10542346]

[52] L. Tsoref, H. Shinar, Y. Seo, U. Eliav, and G. Navon, "Proton double-quantum filtered MRI – A new method for imaging ordered tissues", *Magn. Reson. Med.,* vol. 40, no. 5, pp. 720-726, 1998. [http://dx.doi.org/10.1002/mrm.1910400512] [PMID: 9797155]

[53] R.C. Lyon, J. Pekar, C.T. Moonen, and A.C. McLaughlin, "Double-quantum surface-coil NMR studies of sodium and potassium in the rat brain", *Magn. Reson. Med.,* vol. 18, no. 1, pp. 80-92, 1991. [http://dx.doi.org/10.1002/mrm.1910180110] [PMID: 2062244]

[54] A. Stevens, P. Paschalis, and T. Schleich, "Sodium-23 and potassium-39 nuclear magnetic resonance relaxation in eye lens. Examples of quadrupole ion magnetic relaxation in a crowded protein environment", *Biophys. J.,* vol. 61, no. 5, pp. 1061-1075, 1992. [http://dx.doi.org/10.1016/S0006-3495(92)81916-1] [PMID: 1600073]

[55] G.S. Payne, and P. Styles, "Multiple-quantum-filtered ^{23}Na NMR spectroscopy in model systems", *J. Magn. Reson.,* vol. 95, no. 2, pp. 253-256, 1991.

[56] J.L. Allis, "The application of cation NMR to living systems: ^{87}Rb NMR", *Annu. Rep. NMR Spectrosc.,* vol. 26, pp. 211-246, 1993. [http://dx.doi.org/10.1016/S0066-4103(08)60060-5]

[57] J. Pekar, and J.S. Leigh, "Detection of biexponential relaxation in sodium-23 facilitated by double-quantum filtering", *J. Magn. Reson.,* vol. 69, no. 3, pp. 582-584, 1986.

CHAPTER 2

Dynamic Properties of Bound Water in Natural Polymers as Studied by NMR Relaxation

Abstract: The chapter presents the results of NMR methods in studying natural silk *Bombyx mori* with low water content (0.07 g H_2O per g dry mass). The silk fibers have been also studied on mechanical perturbations measuring the stress-strain dependences for silk fibers. The results obtained are compared with published data on mechanical studies of polymer materials and model calculations of stress-strain curves. The free induction decays (FIDs) in the silk samples have been analyzed within the model of two components: a slow relaxation component was associated with water protons whereas the fast one was related to macromolecular protons. The results discovered a slow molecular mobility and strong interaction of water molecules to silk macromolecules. The chapter presents also NMR relaxation methods in studying interaction of water with macromolecules in oriented and nonoriented collagen fibers with different cross-linking level. The NMR relaxation times (T_1 and T_2) have been studied in the collagen fibers oriented along the static magnetic field B_0. Several NMR relaxation experiments have been done on randomly oriented fibers and for collagen samples with varying hydration level. Cross-relaxation effect has been studied on silk and collagen samples with low water content. Correlation times as characteristics of molecular motions have been considered to compare quantitatively the mobility of water and biopolymer macromolecules in natural silk fibers and collagen tissues with different degrees of binding molecules.

Keywords: Collagen, Correlation times, Cross-relaxation (CR) rate, Elastic deformation, Fourier transform (FT), Free induction decay (FID), Magnetic Resonance Imaging (MRI), Maximal stress, NMR, Second moment, Silk, Spin-lattice (longitudinal) relaxation time T_1, Spin-spin (transverse) relaxation time T_2, Strain ε, T_2 anisotropy effect, The activation energy, Water.

NATURAL SILK *BOMBYX MORI* WITH LOW WATER CONTENT

Many synthetic polymers are copolymers consisting of several monomers. The silk fibers mainly contain polyamino acid-based fibrous proteins [1]. The fibers of silk are biological polymers. The primary sequence and links between the monomers of these biopolymers are responsible for the formation of well-defined structure [1, 2]. NMR methods have been applied in studying hydration properties of silk fibers and silk microstructure [3 - 7]. The information acquired in these

Victor V. Rodin

studies was also important for understanding the intermolecular interactions and properties of other hydrated biopolymers [6 - 11]. The methods of NMR get the information on bound water and characteristics of macromolecules in tissues and fibers [3 - 5, 10 - 15]. A development of the materials based on silk polymers is associated with methods which are able to study hydrated properties of fibrous polymers. In particular, the properties of the silk fibers at such extreme external factors as low water content are important [4 - 9, 16].

It was found that the signal of silk protons after applying 90° pulse can be comprised from the sum of several components (Fig. **1**) [6, 16]. A rapidly relaxing component gave the information about the protons with highly limited mobility whereas a slowly decaying component characterised mobile protons, *i.e.*, water protons. After Fourier transformation of the natural silk FIDs (Fig. **1**) from the time domain to the frequency domain, a characteristic spectrum $f(\omega)$ can be obtained [5 - 7, 16]. In such a spectrum, the narrow resonance line would correspond to the relaxation component with slow decay of FID whereas the fast relaxing component would transform into broad line of the spectrum [16]. The dipole-dipole interaction between protons determines a shape of such a spectrum $f(\omega)$. For the normalized function $f(\omega)$ describing the spectrum with resonance frequency ω_0 (at this frequency the signal intensity has a maximum), the n_{th} moment is introduced according to eq. (1) [6, 13, 16, 17]:

$$M_n = \int_0^{+\infty} f(\omega)(\omega - \omega_0)^n \, d\omega \tag{1}$$

The spectrum is broadened by dipole-dipole nuclear interaction. This is symmetric function relatively to the field frequency ω_0. For such a spectrum, all the odd moments are equal to zero [6, 18]. In practical applications of spectrum moments, only a second moment is taken into account. It does mean that integral in eq. (1) is considered at $n = 2$ [4, 13, 17].

The signal of silk protons and that of water compose the total time-domain signal in *Bombyx mori* silk as $S = S_{\text{w}} + S_{\text{silk}}$ (Fig. **1**). The best fit for the silk part of the FID is the Gaussian-sinc function [5 - 7, 16]. This function, S_{silk}, contains the intensity of silk protons at $t = 0$ as S_{s0} and the parameters a, b, which are needed for the calculation of the second moment M_2 as presented in following expressions (2):

$$S_{silk} = S_{S0} \exp(-a^2 t^2 / 2) \sin c(bt) \qquad M_2 = a^2 + \frac{b^2}{3} \tag{2}$$

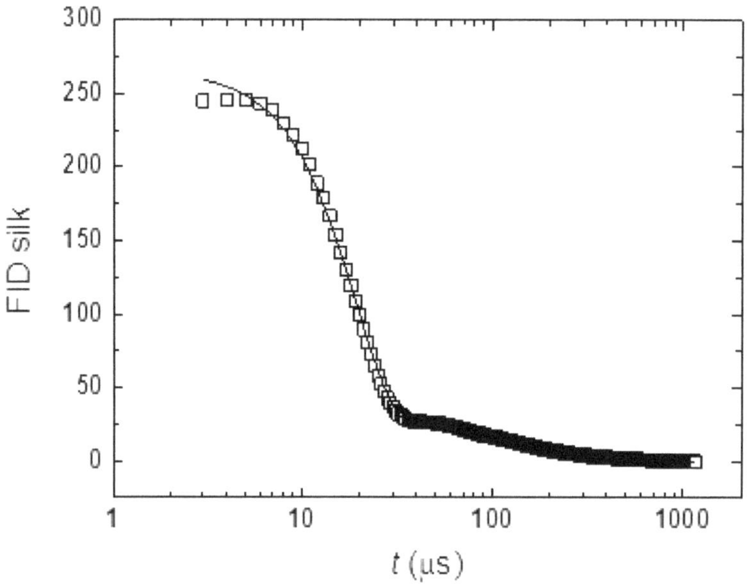

Fig. (1). Proton FID for natural silk *Bombyx mori* sample with HL=0.07 g H_2O per g dry matter measured at *T*=298 K and 20 MHz.
Open Squares: Experimental Data
Solid line: fitting to the expression presenting a sum of exponential and Gaussian-sinc functions [4 - 7, 16].

When the goal of the study is to monitor the water protons, it is necessary to remove a signal of macromolecular protons from the measurable signal. To do this, an additional delay is produced after 90° pulse. After the Fourier transform, NMR spectrum would be recorded then as a single line, *i.e.*, only as resonance signal of water protons [4, 6, 9]. A T_2 value of water protons in the *B.mori* silk with low water content is very short, and this could not be measured by the CPMG pulse method. Then, the line shape of the water resonance signal could be used to estimate the transverse relaxation time [7]. An additional estimation of spin-spin relaxation time T_2 for the water protons of the signal could be done after an isolation of this part from the total measured FID in the sample of natural silk *Bombyx mori*. Then, it is possible to fit the component of slowly relaxing protons by the function [6, 7, 9, 19]:

$$V(t) = S_0 e^{-\pi\omega_L t - \left(\frac{\pi\omega_G t}{2(\ln 2)^{1/2}}\right)^2} \quad \textbf{(3)}$$

In this expression, ω_L and ω_G are the *Lorentz* and *Gauss* widths of the NMR-spectra at half of the peak height in frequency scale, respectively. S_0 is the signal intensity at $t = 0$. T_2 is connected with ω_L according to $T_2 = 1/\pi\omega_L$ [6, 9, 16].

Natural silk is a polymer with large fraction of crystalline domains [20, 21]. The main protein in silk macromolecules is fibroin (~390 kDa [22]). This protein contains glycine (G, 43.7% of all amino acid residues), alanine (A, 28.8%) and serine (S, 11.9%), respectively. In order to form the silk fibroin, two chains (H and L) are used. Disulfide bond connects the H chain (350 kDa) and the L chain (25 kDa) [20]. The crystalline parts of silk are formed by the regular region of the H chain. This chain has the amino acid sequence $(–GAGAGS–)_n$. It is often, this is approximately considered as $(–GA–)_n$ [20]. Silks can serve as accessible block copolymer models for exploring the useful properties of the materials with alternative building blocks. Unique structural organization of silk materials can explain the strength, toughness and stiffness of natural silks [22]. The experimental and computational studies showed how some physical mechanisms control the nonlinear material behaviour of the silk fibers at loading. Those features are due to the protein sequence which generates secondary structures.

Alanine generates very stiff and strong crystals and supplies mechanical integrity. These domains are named *A*-block. These blocks do not mix with water [22]. The regions with rich amount of glycine result in disorganized and soft domains that like to mix with water. These regions are called *B*-block. A combination of *A*- and *B*-blocks leads to silk threads with diverse mechanical properties [22]. These properties are due to the secondary structure of silks.

As to *B.mori* silk, the crystalline areas are stack of sheet structures which are bounded by hydrogen bonds [4, 6, 21]. OH groups of serine bind to the lattice of H-bonds and form a sheet structure. The macromolecular sheet structure has the distortions associated with the presence of serine residues [20, 21]. These silk regions may contain the water molecules. Magnetic resonance imaging (MRI) has also been applied to assess the changes in water state of *B.mori* silk. MRI allows visualization and characterization of water molecules in different porous materials [14, 23 - 26]. Fig. (**2**) shows several images taken on *B.mori* silk fibers (transverse images and the ones along the silk fibers) highlighting the places (white spots) of water excess.

Fig. (**3**) presents the maximal stress σ_{max} in *B.mori* natural silk fibers after heat treatment as a function of the fiber diameter *d*. As for initial fibers, σ_{max} tends to decrease with an increase in the fiber diameter. Silk samples were subjected to heat treatment in a thermal cabinet at 50 and 100 ℃ for 5 and 20 min. All the deformation studies of the initial and heat-treated samples were performed at room temperature [7]. The experiments showed also that the water content of silk fibers remained constant throughout the mechanical tests [4]. The strain ε of the sample was calculated as a percentage in terms of relative elongation $\Delta L/L_0$ of the sample as $\varepsilon=(\Delta L/L_0)\cdot 100\%$ (L_0 is the initial length of a fiber before mechanical test, L_0=10 mm) [7]. In natural samples, the σ_{max} and ε_{max} are considered [27, 28] to determine the maximal stress and maximal strain, respectively, as the elastic limit in the linear elastic region, in which the law $\sigma = E{\cdot}\varepsilon$ is valid (*E* is Young's modulus) [4, 7]. After heat treatment at 100 ℃, ε_{max}=2.40±1.13% and, within the standard deviation, does not differ significantly from the values for the initial fibers.

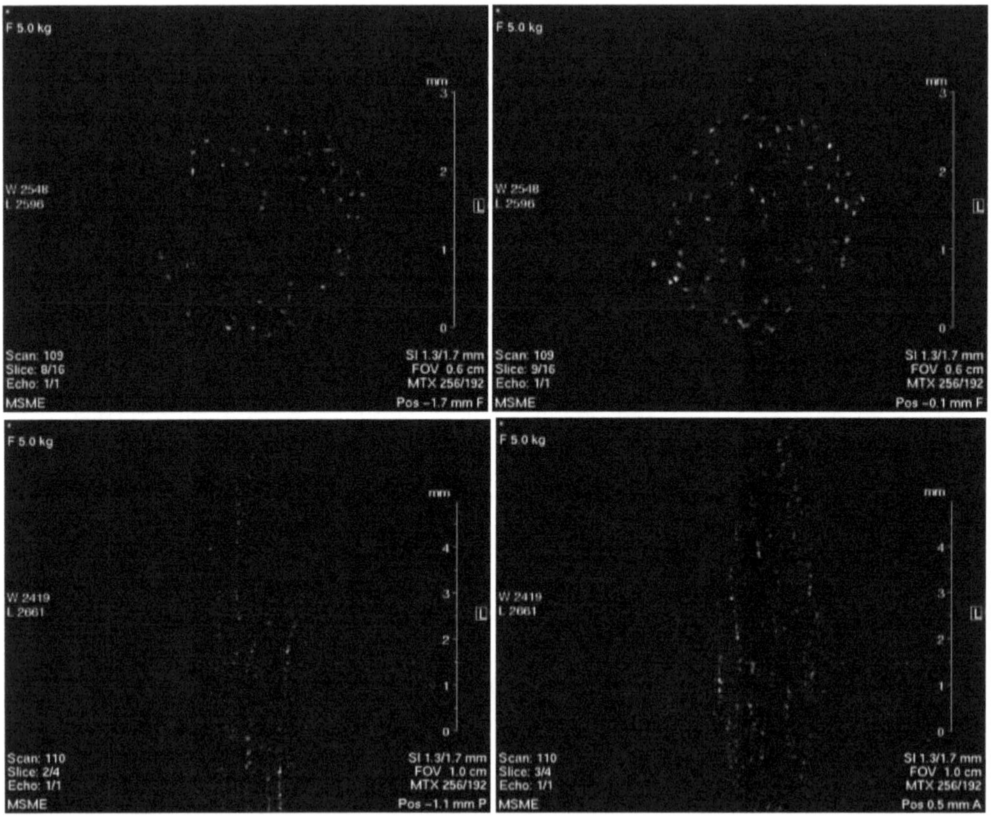

Fig. (2). 400 MHz ^{1}H NMR Images of *Bombyx mori* natural silk fibers. At fixed echo time (*TE*, the delay between the NMR excitation and the collection of the MRI data), the relative intensity of each pixel is characterised by the liquid proton density (M_0) and the local spin-spin relaxation time T_2 giving the intensity of each pixel of an image approximately as: $M = M_0 \exp(-TE/T_2)$.
NMR images are presented for two different slices in transverse (*top*) and longitudinal (*bottom*) planes. The images were acquired using the Multi-Slice Multi-Echo (MSME) image pulse sequence [23 - 26] with the following acquisition parameters: number of slices = 4 (*bottom*) /16 (*top*); slice thickness (ST) = 2.76 mm; inter slice thickness (IT) = 2.91 mm; *TE* = 7.4 ms (top) / 6.2 ms (bottom); recycle delay (RD) = 3.0 s. Each image required about 2 hrs data acquisition.

Fig. (**4**) shows the histogram comparing the maximal stresses σ_{max} in the heat-treated and initial fibers, which are respectively equal to 479±230 MPa (at *d*=21±1 μm) and 1004±132 MPa (at *d*=20±2 μm). In heat-treated fibers with a diameter of d=29±2 μm, σ_{max} is still lower (242±47 MPa). The decrease in σ_{max} demonstrates impairment of the strength characteristics of silk fibers after exposure to heat, which is also accompanied by an approximately twofold decrease in the slope of the linear part of $\sigma=f(\varepsilon)$ curve [4, 7].

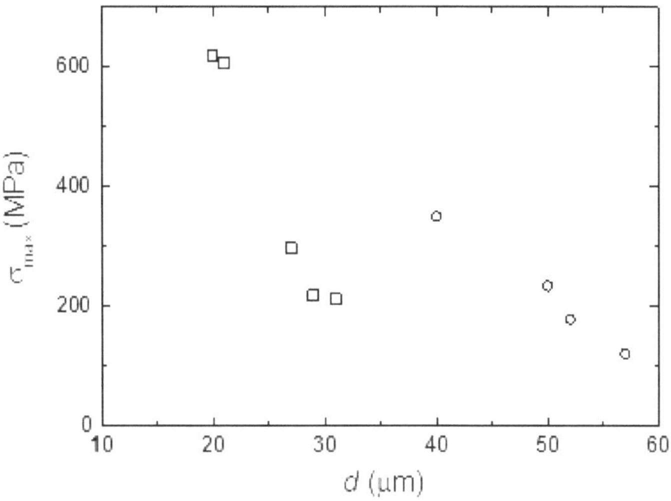

Fig. (3). Maximal stress σ_{max} in *Bombyx mori* natural silk fibers after heat treatment at 50 ℃ (circles) and 100 ℃ (squares) as a function of the fiber diameter.

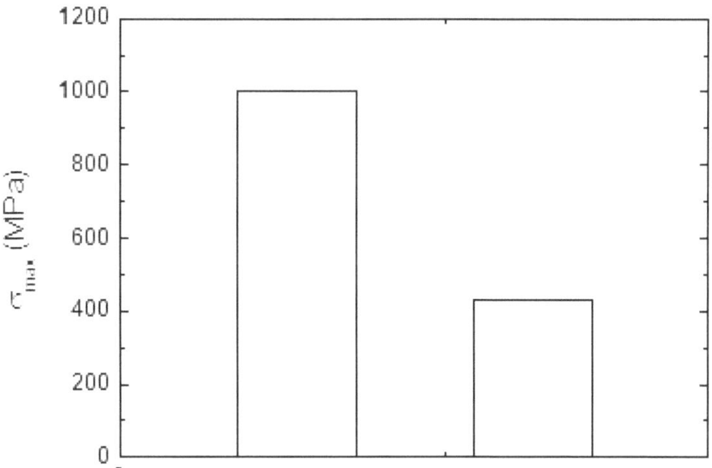

Fig. (4). Comparative histogram of the σ_{max} values (averaged value over five experiments on tension) of initial (*left*, standard deviation ≈132 MPa) and heat-treated (*right*, standard deviation ≈ 210 MPa) *Bombyx mori* natural silk fibers. The fiber diameter d=20±2 μm. The heat treatment was carried out at 100℃ for 20 min.

The mechanical tests for longitudinal tension of initial *B.mori* silk fibers of different diameter showed that within the range of *d*=20-40 μm σ_{max} decreases from 1004±132 MPa (at *d*=20±2 μm) to 189±72 MPa (at *d*=39.5±0.6 μm). These data are consistent with the data of Jelinski *et al.* [29], who also observed that σ_{max} decreases with an increase in the *B.mori* silk fiber diameter and assumed that the large-diameter fibers differ from small-diameter ones by larger number of intermolecular cross-links. The decrease in σ_{max} did mean narrowing the stress range in which Hooke's law for elastic deformation (strain) in uniaxial tension is valid. The same linear region for natural collagen fibers is determined by the stretching the macromolecules in disturbance of cross-links between them [27, 28]. Neighbouring macromolecular chains slip past one another as the fibers are stretched. According to data [27 - 29], this molecular slip is accompanied by considerable stress. Some changes in structure and number of cross-links can lead to narrowing or vanishing the linear region in the $\sigma=f(\varepsilon)$ curve. Comparison between the results on *B. mori* silk [7] and the data on the tension of synthetic Nylon 6 [30] shows that, for these materials, the linear part of the $\sigma=f(\varepsilon)$ curve is observed even at low tensile loads. The σ_{max} values for synthetic Nylon 6 and natural silk prove comparable. Takahashi *et al.* [20] supposed that the structure of *B.mori* natural silk can contain the same unordered regions as those detected in Nylon 6 [30], *i.e.,* silk fibers have the same possibilities of chain slipping which are in Nylon 6. Takahashi [31] also reported that sheet structures formed in Nylon 6 by hydrogen bonds slip past one another at tension.

Heating the *B.mori* silk fibers not only removes residual water but also disturbs other H-bonds between macromolecular chains, which are responsible for the sliding of sheet structures in longitudinal stretching the fibers [7]. A removal of residual water from the silk fibers resulted in a distortion of polymer structure which did not restore the properties after heat action [7]. This is in line with the data that water in natural silk is strongly bound to the protein macromolecules. A self-diffusion of water in the silk *Bombyx mori* samples with HL=0.18 was restricted [3]. At the decrease of HL, a coefficient of water self-diffusion decreased too. When humidity level of *B.mori* silk became equal to HL=0.07 g H_2O per g of dry matter, it was not possible to get echo-attenuation curve, *i.e.*, to measure D_{app} using STE pulse sequence. For the silk with this HL value, the relaxation time T_2 was measured from the width of the NMR spectrum [3 - 5, 7] produced by FT of the time-domain signal into frequency-domain.

Modelling solid-like part of FID in the silk by Gaussian-sinc function resulted in the relaxation constant of 15 μs as a characteristic of mobility for strongly bound protons. The contribution of these protons to the total NMR signal is about 85% [7]. With these data [3, 4, 6, 7, 33], the fast relaxing protons in partially crystalline polymers could be considered as a fraction incorporated in the crystal lattice. The experiment on natural silk *B.mori* resulted in second moment $M_2 = 8.68 \times 10^9$ s^{-2}. This value was comparable to the data $M_2 = 7.5\times10^9$ s^{-2} [17] and M_2=10×10^9 s^{-2} [6, 7, 13 and references therein] obtained on cellulose. The intramolecular contribution $b^2/3$ to the second moment for natural silk is 3.74×10^9 s^{-2}, and the intermolecular contribution is 4.94×10^9 s^{-2}. The intramolecular contribution to M_2 is determined by the interaction of nuclei within rotating groups [7, 18, 33]. At the isotropic rotation of nuclei groups, the intramolecular contribution to M_2 should be zero. As to silk *B.mori*, the intramolecular contribution to M_2 is significant. This supposes an anisotropic motion of water molecules in these fibers. This is in line with other results of NMR methods and the data on the properties of materials [3 - 5, 20, 29, 33, 34]. The data of MacKay *et al.* on timber, cellulose and pectin [13, 17] showed that the second moment was depending on the humidity of the samples. Some studies and analysis of the *B.mori* silk fibers as complex heterogeneous structures [4, 6, 16, 17] predicted that second moment in silk should be much smaller than the M_2 values of a hydrocarbon chain of methylene (CH_2) groups (M_2~20×10^9 s^{-2}). In heterogeneous systems, the experiment yields the measured average second moment, which is used to analyze the change in molecular motion, *e.g.*, in dehydration of the systems [18]. The contribution of the "rigid" component to the second moment is then evaluated under certain assumptions [17, 18]. The second moment in natural silk fibers could be estimated from experimental data, *e.g.*, using one approach [17]. For the estimation of FID signals of a complex system that clearly comprises a "rigid" component and a more flexible component (Fig. **1**) the approach represent the experimental second moment as a combination $M_{2exp} = A_r M_2(1) + (1-A_r)M_2(2)$ of two components, $M_2(1)$ and $M_2(2)$. Here A_r is the fraction of protons related to the "rigid" component, and $M_2(1)$ is its second moment. Since $M_2(1) >> M_2(2)$, the contribution of water component is ignored because its second moment is small, and then the second moment of "rigid" component could be approximated in terms of $M_2(1) = M_{2exp}/A_r$ [17]. Thus, this approach ignores the water component in measured silk sample. This resulted in an estimate of the second moment as M_2 = 10×10^9 s^{-2}. The result obtained testifies that this M_2 value is still substantially smaller than M_2 for CH_2 groups in a hydrocarbon chain [16, 33]. The analysis of the data on cellulose assumes that mobile protons can occur on the surface of microfibrils or in amorphous regions [17]. The M_2 data in *B.mori* silk gives the information about flexible areas and domains with restricted flexibility. It is possible to model these regions to describe the structural properties of fibers. Then, it is assumed that residual water in *B.mori* silk plays a certain role in the change of the fraction of flexible areas [4, 6, 9]. However, the contributions of different components of the FID could be estimated only when the proton densities of water and macromolecules are taken into account [7].

FID signal in *B.mori* silk at HL=0.07 g H_2O per g dry matter has slowly decaying component with spin-spin relaxation time $T_2 \sim 0.21$ ms. This result is comparable with T_2 values for water component in biopolymers at similar humidity level and experimental conditions [6 - 9]. Transverse relaxation rate R_2 of water protons in natural silk *B.mori* can also be analysed in terms of single correlation time model in accordance with eq. (4) and eq. (5) [6 - 8]:

$$R_2 = \frac{2}{3}\eta C\{J(0,\tau_c) + \frac{5}{3}J(\omega_0,\tau_c) + \frac{2}{3}J(2\omega_0,\tau_c)\} + (1-\eta)R_{2w} \tag{4}$$

$$J(\omega,\tau_c) = \frac{\tau_c}{1+\omega^2\tau_c^{\ 2}} \tag{5}$$

Here, η is the fraction of bound water molecules. η shows the ratio of the number of water molecules bound to the biopolymer surface to the total number of water molecules. $C = 2.5 \times 10^{10}$ s^{-2} is the dipolar coupling constant for the protons, and R_{2w} is the spin-spin relaxation rate of protons of pure free water. R_{2w} ~0.5 s^{-1} at T=298 K [35 - 37]. At a water content of 0.07 g H_2O per g dry mass, all of water in natural silk is bound (η=1). Then, eq. (4) and eq. (5) resulted in $\tau_c = 2.7 \times 10^{-7}$ s. The comparable τ_c value, (3.9–4.5) $\times 10^{-7}$, has been obtained for gelatin with a water content of (0.08–0.1) g H_2O per g dry mass at assumption of η=1 [6, 37].

The correlation times of the rigid domains of *Bombyx mori* silk could be estimated according to the distribution $K(\tau)$ of the correlation times. It is considered that this distribution is connected to the experimental transverse magnetization decay curve $F_2(t)$ of polymer protons according to eq. (6) [6, 7, 32]:

$$F_2(t) = \int_0^\infty K(\tau)e^{-t/T_{2m}}d\tau \tag{6}$$

The eq. (7) establishes a relationship between the correlation time τ_c, the second moment M_2, the spin-spin relaxation time T_{2m} of macromolecule protons, and the spectrometer operating frequency ω_0 [6 - 8]:

$$\frac{1}{T_{2m}} = M_2\left[\tau_c + \frac{5}{3}\frac{\tau_c}{(1+\omega_0^2\tau_c^2)} + \frac{2}{3}\frac{\tau_c}{(1+4\omega_0^2\tau_c^2)}\right] \tag{7}$$

The values of M_2 were obtained in the experiment and then were used in the calculation of the correlation times by eq. (7). It was estimated that molecular motions in the rigid domains of *B. mori* silk have correlation times between 2.9 × 10^{-6} s and 7.5 × 10^{-6} s.

Thus, NMR approaches based on investigating the second moments and modelling the FID signals can clarify the details of deformation influence in natural silk *B.mori*. This allows one to reveal specific features of the interaction of small amounts of water with macromolecules including the changes in the structural characteristics of fibers under external influence. An estimation of the correlation times of molecular motions enables one to compare quantitatively the mobility of water and biopolymer macromolecules in natural silk *B.mori* samples with different level of binding molecules. This is essential to understand the processes of dehydration of natural polymers. Water in the *B. mori* silk samples studied is strongly bound to fibroin macromolecules, and the removal of this water by heat treatment does not lead to restoration of the initial deformation characteristics of fibers [7].

NMR RELAXATION IN NONORIENTED AND ORIENTED COLLAGEN FIBERS

Different NMR methods have been applied in studying collagens and collagen based tissues [3, 16, 35 - 57]. The collagen macromolecules are composed from the triple helices. The hydrogen bonds between two CO groups or between CO and NH groups stabilize the chains of these helices [38 - 40]. Water plays an important role in the stabilization of the structure of collagen macromolecules. The study of the intermolecular interactions and water mobility near the collagen macromolecules is important for clarifying the properties of biomaterials [35, 37 - 39]. Molecular structure of collagen is highly anisotropic. Therefore, water absorbed in collagen has anisotropic mobility [6, 35, 39 - 43]. The studies of different collagens provide additional information for understanding water-macromolecule interactions related to some diseases of connective tissues [6, 14, 15, 25, 35, 41 - 46]. On the base of these studies, a clarification of relaxation mechanisms in both collagens and other fibrous materials can be done [5 - 9, 35 - 42].

Next section considers the results of NMR relaxation studies of ECSD collagen (from egg cases of lesser spotted dogfish) at HL≥0.6 g H_2O per g dry matter.

NMR Relaxation in ECSD (Randomly Oriented) Collagen Samples

The recovery of longitudinal magnetization for water protons of ECSD collagen at HL≥0.6 was characterised by single T_1 value. The spin-lattice relaxation rate R_1 depended on HL of collagen according to the data of Fig. (**5**). In contrast to T_1, the

CPMG studies showed two exponential components with characteristic times of T_2' = 4.0–4.6 ms (for slowly relaxing component), and T_2" = 1–2 ms (for the fast relaxing component).

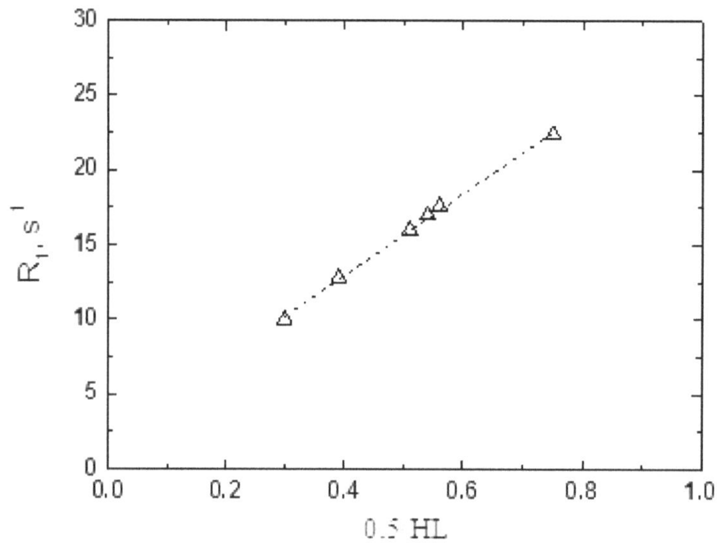

Fig. (5). Dependence of R_1=1/T_1 on water content in ECSD collagen samples [8]. NMR measurements were carried out at 20 MHz and *T*=298 K.

The published results testified [6, 35, 38, 40, 47] that for the fractions of bound (*b*) and free (*f*) water, one can write an expression (8) giving the relationship with the relative number of bound water molecules η:

$$\frac{1}{T_i} = \frac{\eta}{T_{ib}} + \frac{1-\eta}{T_{if}} \quad \mathbf{(8)}$$

$T_{1f} = T_{2f}$ for free water. However, for bound water $T_{2b} << T_{1b}$ [5, 6, 16, 35]. The proton relaxation rates R_2 and R_1 could be considered with the aid of the single correlation time model, *e.g.*, eqs. (4), (5) for R_2, and eq. (9) for R_1 [3, 4, 6, 40, 48, 49]:

$$R_1 = \frac{2}{3}\eta C\{J(\omega_0, \tau_c) + 4J(2\omega_0, \tau_c)\} + (1-\eta)R_{1w} \quad \mathbf{(9)}$$

Here, $C = 2.5 \times 10^{10}$ s^{-2} is the dipolar coupling constant for the protons of water, R_{1w} is the longitudinal relaxation rate of protons of free water (R_{1w} ~ 0.5 s^{-1} at room temperature), and ω_0 is the proton *Larmor* frequency [4, 6, 35]. When water content is less than 0.2 g H_2O per g dry mass, then it is taken $\eta = 1$ in eq. (4) and eq. (9) [37, 40]. In ECSD collagen, HL $\geq$ 0.6 g H_2O per g dry mass. Then, for these HL values, one approach has been considered [6, 40]. According to this [6, 35, 37, 40], the correlation times τ_c have been initially estimated in eq. (4) and eq. (9) ignoring the term containing $(1-\eta)$ and using the ratio T_1/T_2. The same approach estimated the T_1/T_2' ratio for other biological samples with a HL $\leq$ 0.7 g H_2O per g dry mass. In such a case, transverse magnetization decay was described by two exponential curve, for which the slow time constant T_2' and the fast time constant T_2" have been calculated. For ECSD collagen with an HL between 0.64 and 1.42, the same approach allowed the estimation of τ_c as (0.34–0.40) $\times 10^{-7}$ s [6, 8]. The estimated τ_c values and experimental T_1 data for each HL value in ECSD collagen have been used in calculation of η according to the eq. (9). Fig. (**6**) shows the bound water fraction as a function of reciprocal value of HL.

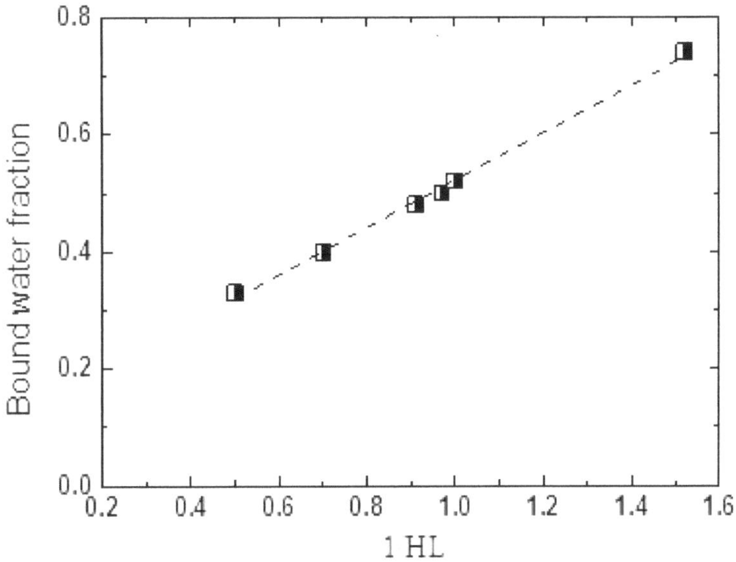

Fig. (6). Bound water fraction η *vs* 1/HL in ECSD collagens [8]. NMR measurements were carried out at 20 MHz and T=298 K

According to the results obtained [8, 35, 40], the relative number of bound water molecules can be expressed by the relationship $\eta \approx C_{wa}/\text{HL}$. The value of C_{wa} is comparable to the water content c_a at which water molecules are strongly bound to biopolymer (c_a=0.5 g/g in the function $\eta = c_a/\text{HL}$ [6, 40]). The relative number of bound water molecules is depending on the humidity level of collagen. For instance, the published data showed [6, 47, 49, 50] that if the collagen samples have a humidity level ≤ 0.5, then the water does not freeze at temperature between 213 and 273 K. If water amount exceeds 0.5–0.54 g H_2O per g collagen, then below 273 K an excess of water can be frozen. The water amount that stabilizes the collagen helices is considered as hydration water [8, 47, 50]. It does not freeze even in the temperature range between 213 and 223 K. This is called as non-freezing (hydrated) water. It was shown in the studies of hydrated collagens that the proton longitudinal relaxation rate of hydrated water (R_{1h}) at humidity level of the sample > 0.5 is connected with measured longitudinal relaxation rate R_1 according to eq. (10) [6, 35, 50]:

$$R_1 2HL = R_{1h} + (2HL - 1)R_{1w} \qquad \textbf{(10)}$$

The NMR studies of randomly oriented ECSD collagens (Fig. **5** and eq. (10)) resulted in estimates of $R_{1h} \approx 27.2\ s^{-1}$ and $R_{1w} \approx 3.9\ s^{-1}$. The spin-lattice relaxation rate $1/T_1$ increased with increasing η. Thus, at decreasing water content from 1.42 to 0.64 the mobility of macromolecular chains decreases. At the same time, the fibril binding energy of collagen increases [35, 50]. With an increase in water content the relative fraction of bound water decreases. This results in destabilising the triple helix. When an amount of water in collagen sample increases, the collagen structure strongly changes. The fibers swell leading to a more free motion of polypeptide chains.

Additionally to the estimation of τ_c values with the aid of T_1/T_2 ratio, τ_c was also calculated from the slow T_2 component using eq. (4) and eq. (5) [8]. These estimates showed that for humidity level range between 0.6 and 1.42 these values of $\tau_c = (0.6–0.8) \times 10^{-8}$ s are already smaller than those obtained on the base of T_1/T_2. Similar difference in correlation times estimated by these two approaches was obtained in the Ref. [37]. The authors explained that result by the fact that at studied humidity level HL ≤ 0.7 spin-lattice relaxation was also affected by dipolar cross-relaxation.

Studying different polymers with water content showed noticed effect of cross-relaxation at HL ≤ 0.6 [5, 6, 35 - 38, 48, 51]. It was found, that the water cross-relaxation rate k in *B.mori* silk fibers (HL = 0.18) is 167 s^{-1} [5]. In collagen from skin of calf (HL = 0.27), $k \approx 170\ s^{-1}$, whereas in collagen from rat-tail tendon (HL = 0.36), $k \approx 40\ s^{-1}$. In chick muscle at HL = 0.78, $k \approx 3.3\ s^{-1}$ [5, 6, 38, 51]. Ref. [37] showed that the cross-relaxation rate decreased in gelatin samples at the increase of HL. In particular, at water content in gelatin ≈ 0.18 g H_2O per g of dry mass, cross-relaxation rate k was $\approx 20\ s^{-1}$. At HL ≈ 0.38, cross-relaxation rate k was $\approx 6\ s^{-1}$, and at HL=0.6, cross-relaxation rate $k \approx 4\ s^{-1}$. The k value began to decrease very fast at HL > 0.20 [37]. The rate k is considered as $k = P\sigma$, where P is the relative number of protons of bound water, and σ is the cross-relaxation rate in a pair of protons. One proton belongs to the rigid biopolymer chain. The σ value characterizes the rate of the transfer of spin energy through the water-macromolecule interface. It is related to the correlation time as shown in eq. (11), where r_{ij} is the distance between the bound water proton and the collagen proton [3, 4, 48, 51]:

$$\sigma = \frac{\gamma^4 h^2}{40\pi^2 r_{ij}^6}[\tau_c - \frac{6\tau_c}{1+4\omega^2\tau^2}] \tag{11}$$

The shortest distance between the bound water proton and the macromolecule proton is 2.5 Å [6, 48, 52]. Based on this r_{ij} value for τ_c presented above, it was estimated that σ = 7–8 s^{-1} [8]. The cross-relaxation causes the bi-exponential behaviour of the longitudinal magnetization recovery. In this case, the exchange rate of spin energy through interface of the water-macromolecule exceeds the spin-lattice relaxation rate of bound water [51]. For ECSD collagen the longitudinal magnetization recovery is monoexponential [6, 8]. Thus, according to the presented estimates, $R_{1h} \geq \sigma$, and the effect of the cross-relaxation is insignificant in studied samples of ECSD collagen. For ECSD collagen, the correlation time τ_h of intramolecular interaction in hydrated water can be calculated on the base of the rate R_{1h} according to eq. (12) [48, 51]:

$$R_{1h} = \frac{3\gamma^4 h^2}{40\pi^2 r_{ii}^6}[\frac{\tau_h}{1+\omega_0^2\tau_h^2} + \frac{4\tau_h}{1+4\omega_0^2\tau_h^2}] \tag{12}$$

Here, r_{ii} is the distance between the bound water protons in water molecule [48, 51, 53]. For the distance r_{ii} =1.58 Å [53], the correlation time related to the dipole interaction of protons of structured and adsorbed water was estimated as $\tau_h = 5.2 \times 10^{-7}$ s. This value of τ_h was comparable to the correlation time $\tau_c = (4.5–7.0)\times 10^{-7}$ s found for bound water ($\eta = 1$) in hydrated gelatin at HL ≤ 0.1 [37].

The results of studying ECSD collagen [8] have been comparable with the findings of Blinc *et al.* [40] and could be explained in the terms of three component model considering free, adsorbed, and structured water. As shown in the Refs. [8, 35, 37, 47], a recovery of longitudinal magnetization is characterised by single T_1 constant because of fast exchange between all the water fractions. However, on the T_2 time scale there is a slow exchange between structured and free water [40]. There was also slow exchange between the fractions of structured and adsorbed water on the time scale for T_2. Besides the Blinc model [40], there are other approaches to compare the NMR relaxation data on several components of T_2 [38, 50, 53 - 55]. The studies of proton transverse relaxation in pig tendon at

HL = 1.5 revealed four T_2 components [38]. Three components in T_2 studies of bovine collagen with humidity level of 1.7 have been discovered [55]. For the cow and calf muscle tissues with HL = 4, two T_2 components have been registered [50]. The studies [38, 50, 53 - 55] were carried out at the same proton resonance frequency as it was done in the investigation of ECSD collagen (20 MHz) [8]. The collagen samples were from different tissues, and they had different water content. There is a plenty of collagen proteins indeed with various features [50 - 56, 58, 59]. For instance, type I collagen forms thick fibrils, but type III collagen forms only thin fibrils. Some published findings informed [57] that collagen of I type has very compact structure which leads to a less efficient interaction of a macromolecule with water. In the tissues with higher concentration of collagen I (in comparison with type III collagen), the values of T_2 for water protons are also measured as increased. The components of T_2'=4.0–6.2 ms and T_2"=1–2 ms recorded for ECSD collagen [8] were comparable with 4.2–7.2 ms and 0.85–1.8 ms, which are two fast relaxing T_2 components from four T_2 components measured in pig tendon collagen with HL = 1.5 [38]. For the ECSD collagen at HL ≥ 1.96 [8], two slowly relaxing T_2 components (~ 98 ms and ~ 15 ms) comparable to slow T_2 components in pig tendon (~ 90 ms and ~ 17 ms) [38] have been registered [6]. The fast relaxing T_2 component (1–2 ms) in the ECSD collagen had a contribution to the total signal of transverse magnetization which decreased with an increase of water content from HL = 0.64 to HL = 1.5. At high humidity level (HL ≥ 1.97), this contribution was not detected. Thus, for the ECSD collagen with HL = 1.97, only three T_2 components (~98, ~15 and ~5 ms) could be detected in CPMG experiment. At HL = 3 g H_2O per g dried matter of ECSD collagen, two slowest of the three detected T_2 components (~250, ~35 and ~7.2 ms) become comparable to two T_2 components registered in collagen-containing tissues of cow and calf at HL = 4 (~220–400 and ~25–35 ms) [57]. The component of water protons in ECSD collagen with T_2=1–2 ms [8] was responsible for the effect of intra- and intermolecular dipole interactions. It was also responsible for the proton exchange on the macromolecule surface [53, 55]. The transverse relaxation rate R_2=1/T_2 of this component could be analysed using the following eq. (13) [8, 53, 60, 61]:

$$R_2 \approx \frac{1}{2\tau_{2exc}} + \frac{9\gamma^4 h^2}{80\pi^2}\tau_{2d}\left[\frac{1}{r_{ii}^6} + \frac{5}{9r_{ij}^6}\right] \qquad \textbf{(13)}$$

Here, r_{ii} and r_{ij} are the same distances as those in eq. (12) and eq. (11), respectively. The spin-spin relaxation rate R_2=1/T_2 is determined by the exchange time τ_{2exc} and the correlation time τ_{2d} for dipole-dipole interaction. For r_{ij} = 2.5 Å [52] and r_{ii} = 1.58 Å [48, 51 - 53], approximation of R_2 data for the fast relaxing component by eq. (13) resulted in the estimates of τ_{2exc} = (1.1–10.0)×10^{-6} s and

τ_{2d} =(2.6–26.2)×10^{-9} s, respectively. At the calculation of R_2=1/T_2 by eq.(13), τ_{2exc} and τ_{2d} were considered as two parameters with contribution of each no less than 5%. The calculated τ_{2exc} for the ECSD collagen was similar to τ_{2exc}=1–3 μs obtained in the Ref. [53] also from eq. (13) for the collagen from cow tissues of different ages. The study of collagen samples in the range of HL = 0.15–0.33 [53] did not show any clear dependence of the τ_{2exc} on the water content. The results on the ECSD collagen [8] informed that the τ_{2d} values are at least an order of magnitude smaller than the values of τ_{2d} = 200–300 ns obtained for 3-month-old calf and 18-month-old bull at HL = 0.15–0.33 [53]. The data of Ref. [53] were calculated at r_{ij}= 2.2 Å. The τ_{2d} values decreased by a factor of 1.5 with an increase of HL value from 0.15 to 0.33 [53]. There was no data obtained at water content HL > 0.33. However, discovered tendency suggests that at HL > 0.33, the τ_{2d} values can decrease further, *i.e.*, at HL ≥ 0.6, it could be already comparable with τ_{2d} values estimated for ECSD collagen at HL ≥ 0.6 [8].

Thus, NMR approaches and models describing the detected proton fractions in the collagen-water systems allow one to reveal specific features of interactions of small amounts of water with biopolymers and to evaluate the possible effects of intra- and intermolecular interactions in calculations of correlation times τ_c as the characteristics of molecular motion.

NMR Relaxation in Oriented Collagen Fibers

Many studies of hydrated collagen tissues showed [8, 35 - 38, 50, 55, 59, 62] that a spin-lattice relaxation of water protons can be characterized by a single T_1. Other T_1 studies, *e.g.*, of oriented C8y and C15m collagen fibers at HL = 0.6 g H_2O per g dry mass showed that the best fit for magnetization curve in the inversion recovery experiment was for a sum of two exponential functions [35, 36]. For C15m collagen, the T_1 components were T_{1long} = (571.7 ± 5.7) ms (population is 82%) and T_{1short} = (24.3 ± 1.6) ms (population is 18%). For C8y, the results were as follows: T_{1long} = (594.0 ± 2.6) ms had the population of 83.7% whereas T_{1short} = (27.0 ± 2.0) ms had the population of 16.3%. The T_1 studies of nonoriented (randomly distributed) fibers showed that both measured T_1 components were similar to T_1 values recorded for the oriented fibers. The same distribution of proton populations (*i.e.,* 84% of all measured protons were associated with long component and 16% with short component) has been obtained for both randomly oriented C15m and C8y samples. No any orientation dependence for T_1 has been discovered [35]. The T_1 studies of a spin-lattice relaxation for oriented collagens C15m and C8y have been compared also at different temperatures. In particular, the T_1 results have been obtained for C15m at 278 K (T_{1long} = 482 ms (82% of all population) and T_{1short} = 13 ms (18%)) and at 263 K (T_{1long} = 419 ms (82% of all population) and T_{1short} = 13 ms (18%)) to add to the data at 298 K [35]. The spin-lattice relaxation rate R_{1slow} ($1/T_{1long}$) increased with decreasing temperature. Water population A_{slow} for this component remained unchanged (circa 82% of all measured signal). Arrhenius plot (ln(R_{1slow}) as a function of 1/T K) for C15m collagen at HL=0.6 resulted in the activation energy E_a=6.18 kJ/mol. T_1 and T_2 have been also studied at different values of hydration level of C15m collagen fibers (from HL=0.45 to 0.87 g H_2O per g collagen) [35]. The results obtained showed that the values of R_1 both for slow and for fast relaxing components decreased with increasing HL.

The extraction of collagen from the tissue and its purification from noncollagen fraction is essential to prepare the material before starting NMR measurement [36, 55, 63].The collagen properties can differ (*i.e.,* the NMR experiments can result in different findings) for collagens of different purification. For instance, the samples of C15m and C8y collagen contained tissues were about 65% of collagen content. The model describing the influence of additional components in tissue (*e.g*, noncollagen proteins, impurities, *etc*.) could be considered also at the analysis of C15m and C8y T_1 data. It is known [38, 55, 62 - 64] that water molecules in collagen fibers are associated with the hydration of collagen. However, the presence of other proteins changes the compartmentalization of water and the distribution of water molecules in hydrated collagen.

The T_1 component with population ~82–84% of all measurable protons is due to the hydration of collagen molecules in fibers [35, 36]. The slow relaxing T_1 component was characterized by big value of T_1 constant, which could be considered as the result of fast exchange between different possible states of water from inside and hydration layer of collagen triple helix. For the analysis of observed slow relaxing component of water protons in collagen fibers, we can apply the approach which was already probed in literature [37, 40]. In particular, the registration of single component for spin-lattice relaxation rate (R_1= $1/T_1$) in hydration layer of collagen with water content higher than 0.5 g H_2O per g collagen could be explained according to eq. (10) [50]. For example, the ECSD collagen study showed [6, 8] that R_{1w} and R_{1h} are applied to the “freezable” water fraction and hydrated one. The fit of R_{1slow} data on C15m fibers with relation of $R_{1slow} = (0.5/\mathrm{HL})\cdot R_{1h} + (1–0.5/\mathrm{HL})\cdot R_{1w}$ at water content HL > 0.5 g H_2O per g collagen yields R_{1w} = 1.04 [35, 50]. The increase of R_{1slow} with decreasing water content is responsible for a decrease of chain mobility and increase of fibrillar packing. The observed tendency is comparable with that in the publications [50, 55, 62]. Only single component in R_1 has been observed in conditions, when one used more pure collagen fibers (85% collagen content) [50, 55]. It is possible to suppose that different experimental conditions (about 65% of collagen content and 35% of noncollagen protein and polysaccharides fractions) result in the observation of water bound on the noncollagen additives as short T_1 component in the C15m and C8y collagens [35]. An exchange between two observed fractions of water molecules associated with two components is not fast on the T_1 scale. As the difference in T_1 value for slow component in C8y (594 ms) and C15m (572 ms) is not big, the water molecules (associated with noncollagen fraction) in C8y and C15m can have comparable mobility with $R_{1fast} = 37\ s^{-1}$ and $R_{1fast} = 41\ s^{-1}$, respectively.

The use of water protons (R_{1w}) and noncollagen protein protons (R_{1ncp}) for modelling R_{1fast} in C15m and C8y samples can be based on the data of Refs. [35, 50]. So, the observed R_{1fast} could be described then according to the equation: $R_{1fast} = p_w \cdot R_{1w} + p_{ncp} \cdot R_{1ncp}$ or (in the case of $R_{1w} << R_{1ncp}$) $R_{1fast} = R_{1w} + p_{ncp} \cdot R_{1ncp}$, where p_w and p_{ncp} are the proton populations of these phases and $p_w + p_{ncp} = 1$. With the substitution of p_{ncp} on $(1–x \cdot \text{HL})$ it is possible to get an estimation of water fraction $(x \cdot \text{HL})$ on the whole protein system, which can be involved in the hydration of noncollagen protein fraction. Using the equation of $R_{1fast} = R_{1w} + (1–x \cdot \text{HL}) \cdot R_{1ncp}$ to fit the R_{1fast} from experimental data [6], it was obtained that $x = 0.41$, $R_{1ncp} = 50.8\ s^{-1}$, and $R_{1w} = 1.01\ s^{-1}$. Thus, $R_{1ncp} = 50.8\ s^{-1} >> R_{1w} = 1.01\ s^{-1}$. These data, R_{1w}, are in line with the values of $R_{1w} = 1.04\ s^{-1}$ from the fitting equation (10) for R_{1slow} [50].

Refs. [6, 36] considered the data on T_1 measurements in C15m and C8y collagen fibers with increasing duration of 180° pulse from 12 to 80 μs. The population of the slow component decreased with RF pulse length, and the population of the fast component increased. The results obtained were in line with the behaviour of longitudinal relaxation in tendon and collagen with different water content [35, 38, 53, 64]. On the base of the T_1 data of these publications, it was possible to consider that the fast T_1 component is due to the effect of collagen protons on the water phase because of the cross-relaxation. It was shown that the protons of macromolecules participate in transfer of magnetization [6, 8, 35 - 37, 55]. To get more details on the cross-relaxation, the GS experiments on C15m and C8y collagen fibers at HL = 0.6 have been carried out [35, 36]. Fig. (**6**) of Chapter 1 shows the results of studying proton signal of water in C8y collagen at HL=0.6 applying *Goldman-Shen* pulse sequence. The equations (7) - (12) of Chapter 1 were fitted to the experimental data varying the parameters ω_L, ω_G to calculate the transfer rates of magnetization k_w and k_c, the spin-lattice relaxation rates R_{1w} and R_{1c}, and the populations of proton fractions [35]. In particular, for C8y collagen fibers, the parameters obtained from the fitting were $k_w = 18\ s^{-1}$, $k_c = 23\ s^{-1}$, $R_{1w} = 1.65\ s^{-1}$, $R_{1c} = 1.38\ s^{-1}$, whereas for C15m fibers, $k_w = 19.5\ s^{-1}$, $k_c = 22\ s^{-1}$, $R_{1w} = 1.75\ s^{-1}$, and $R_{1c} = 1.52\ s^{-1}$ [36]. At comparison of the GS data for C15m and C8y collagens, it was discovered that the cross-relaxation phenomenon is responsible for the longitudinal relaxation of water protons in both types of hydrated fibers. The effect results in registration of two T_1 components. The cross-relaxation process between the phase of macromolecule protons and the phase of water protons can be considered with the aid of three steps. First, there is a transfer of spin energy from water protons to bound water. Moreover, the transfer of spin energy further from protons of bound water to protons of macromolecule surface and the transfer of spin energy from protons on collagen surface to inner protons of the biomacromolecules [5, 6, 35, 36, 53].

The T_2 studies of C15m and C8y collagen fibers discovered that a nonexponential decay may be described by two components of spin-spin relaxation time T_2: T_{2short} and T_{2long} ($T_{2short} < T_{2long}$) [35]. These two T_2 values were dependent on the pulse spacing separation (in CPMG sequence) and changed in both C8y and C15m collagens from 6.7 ms to 15.8 ms for long component T_{2long} and from 1.5 to 3.95 ms for short one T_{2short}. The comparison of these T_2 results with published data on tendon, cartilage, and other collagen tissues [55, 59, 60, 64] showed that those studies have been carried out with the samples of HL = 0.8–1.7 g H_2O per g dry mass whereas the C8y and C15m collagens had HL= 0.6. Those publications did not perform also the detail T_2 CPMG pulse spacing (t_{cp}) study [38, 55, 59, 63, 64]. Fung *et al.* studied T_1 and T_2 in hydrated collagen [64]. However, within the T_2 study, they only showed that CPMG curves depend on pulse spacing and presented results for t_{cp} = 1 ms and t_{cp} = 2 ms (at HL = 1.14) without details and analysis. Peto *et al.* [38] studied tendon and observed four T_2 components. The authors showed only that the fast relaxation component depended on the pulse separation in a CPMG study, *i.e.*, this T_2 relaxation time increased at decreasing t_{cp} from 150 μs to 100 μs. They attributed this component to mobile protons of macromolecule. To explain the data on transverse relaxation, the authors [38] described some possible reasons of obtained results. For example, at consideration of diffusion through the static field inhomogeneities, they estimated the static field inhomogeneity of the NMR spectrometer as 1 G/cm. They did the estimation of diffusion coefficient to explain a character of T_2 changes with diffusion distance and t_{cp} in CPMG experiment. However, this resulted in unreasonable value. Then, the authors did a conclusion that molecular diffusion is negligible in their case. Moreover, they considered the proton exchange between sites separated by a chemical shift [38], compared their T_2 results for tightly bound protons with other data [63 - 68], and concluded that considered reason cannot also explain the data. However, other possible mechanism could be associated with the T_2 data dependence at short t_{cp} values. This mechanism results in decay of transverse magnetization which does not define any more T_2. It results in the measurement of relaxation time in the rotating frame ($T_{1\rho}$) at spin-locking frequency ν_{sl}= 1/(4·t_{cp}) [38, 65].

The CPMG experiments with variable t_{cp} values on mouse muscle tissue discovered a spin-locking effect and showed that at very short pulse spacing time there was the measurement of $T_{1\rho}$ instead of T_2 indeed [65]. Three T_2 components have been registered in muscle tissues. According to these multiexponential CPMG results in muscle tissue, the fast relaxing T_2 component did not change the value on the pulse separation at the sufficiently short t_{cp}. At the same time, the long T_2 components was depending on a pulse spacing [65]. So, a similar spin-locking effect can be considered for collagen C15m and C8y studies in the range of t_{cp} from 200 µs to 50 µs. In this range, *i.e.*, at a decrease of t_{cp} to 50 µs, the T_2 values increased. For comparison, the CPMG study of doped water showed that in the range of t_{cp} from 50 µs to1 ms there was no change in the T_2 value [35]. The studies of oriented fibers of C15m samples with different values of hydration level (0.45, 0.58, 0.6, 0.71, and 0.87) and randomly oriented C15m fibers showed the increase of T_2 values with increasing t_{cp} from 300 µs to 1ms.

In randomly oriented C15m fibers, R_{2fast} and R_{2slow} did not increase with the increase of t_{cp} from 50 µs to 200 µs. The population of slow component A_{slow} in these fibers increases from 75% to 98%, and the population A_{fast} decreases from 25% to 2% at increase of t_{cp} from 50 µs to 300 µs [35]. The study of spin-locking effect in mouse muscle registered three components in CPMG experiment and discovered a variation in T_2 value for "medium" (35 ms) and "long" (75 ms) components only [65]. At increase of t_{cp}, the relaxation times of "medium" and "long" components decreased with the degree of spin-locking. At the same time, the short relaxation component disappeared. Similarly, short T_2 component in both C8y and C15m collagens at HL = 0.6 disappeared with increasing t_{cp} to 200-300 µs [35]. In Ref. [38], the short T_2 component disappeared when $t_{cp} \geq 500$ µs. It was shown that the population of each component was dependent on the change of t_{cp} [65]. The studies of oriented C8y and C15m collagen fibers showed that the fractional weightings of both T_2 components were also dependent on t_{cp}. However, in the samples of randomly oriented C8y collagen fibres, the populations of the components were approximately constant in the range of t_{cp} = 50–300 µs, and it changed only when long time of interpulse spacing t_{cp} exceeds 500 µs [35]. This is due to a change in the multiexponential character of the fit [38, 65]. The CPMG measurements at longer t_{cp} could result in the appearance of two T_2 components in C8y and C15m fibers again, when multiexponential character of a fit gives increasing values of T_2. At change of t_{cp} value from 1 ms to 300 µs, spin-locking effect doesn't influence on the T_2 values.

For randomly oriented C8y and C15m collagen fibers, only single component was observed for all values of $t_{cp} > 300$ µs. On the base of these results, it was possible to suppose that in hydrated C15m and C8y collagen fibers, there are very slow proton motions that modulate dipolar interactions with typical low-frequency $T_{1\rho}$

dispersion [65]. At varying water content in oriented C15m fibers (*e.g.*, at HL = 0.45, 0.58, 0.6, 0.71 and 0.87 g H_2O per g dry matter), a dependence of T_2 on t_{cp} showed similar tendency (at short t_{cp} values) [35]. However, from CPMG curves for C15m at HL = 0.45, it was seen that spin-locking effect was weaker than that at HL = 0.6 and HL = 0.87. At HL = 0.45, a short T_2 component disappeared when $t_{cp} \geq 200$ μs. At HL=0.6, the disappearance of short T_2 component was noticed at longer t_{cp} values (≥ 300 μs), whereas for C15m collagen with water content of HL = 0.87, the short T_2 component disappeared at $t_{cp} = 500$ μs [35].

In oriented samples (along the static magnetic field), collagen fibers are aligned at angle of $\theta \approx 0°$ with respect to static magnetic field B_0. However, in the samples with randomly oriented fibers, different angles can be in the slope of the fibers to the magnetic field. Anisotropic arrangement of collagen macromolecules results in anisotropic motion of bound water. Thus, the T_2 anisotropy can be observed in CPMG studies of water in collagenous tissues [38, 62, 67 - 70]. When CPMG measurements are performed on oriented and randomly oriented collagen fibers, anisotropy of transverse relaxation of water protons can be clarified.

In order to observe distinct anisotropic dependence of transverse relaxation [38, 62, 68 - 71], the angles of $\theta = 0°$, $\theta = 55°$, and $\theta = 90°$ were studied. Takamiya *et al.* discovered that in normal tendon, the T_2 relaxation of water is bi-exponential [69]. An anisotropy for both T_2 components (different T_2 values for various angles of θ) was observed. Thus, T_2 relaxation was depended on orientation of collagen fibers with respect to direction of magnetic field B_0. According to Peto *et al.* [38], the largest T_2 values were obtained in the experiments at $\theta = 55°$, *i.e.*, at 'magic angle'. In the experiment with this angle, the line between two spins makes the angle of 55° with the direction of the static magnetic field B_0. When the dipole-dipole interaction is decoupled, the spin-spin contribution to relaxation is diminished, and T_2 is increased [62]. At $\theta = 0°$, the value of T_2 was the smallest [38].

When the C8y and the C15m collagen fibers are distributed randomly, some fibers are in orientation of $\theta = 55°$. Because of randomly oriented fibers in such a sample, some fibers are at other angles $\theta \geq 0°$. Then, in average, T_2 values in the sample with randomly distributed fibers exceed the ones for fibers with orientation of $\theta = 0°$. Thus, the sum effect of T_2 anisotropy at randomly mixed collagen fibers (with different orientation to magnetic field B_0) results in increased values of T_2 in comparison with the fibers of only one orientation (at $\theta = 0°$). Considering spin-locking effect in randomly oriented fibers, it is necessary to emphasize that dependence of $T_2 = f(t_{cp})$ may be different for different T_2 components. In the case of mixture of fibers with various orientations to static magnetic field, *i.e*, in randomly oriented fibers, some orientations can have T_2

components with weak dependence of $T_2 = f(t_{cp})$.

T_2 may not change with increase of t_{cp}. So, it results in weak spin-locking effect or total absence of such effect. Additional T_2 studies on C15m and C8y collagens with *Carr–Purcell–Freeman–Hill* (CPFH) pulse sequence [72] showed two T_2 components (6–7 ms and 2–3 ms) without strong decrease / increase with t_{cp} pulse spacing as it was before in CPMG experiments [35]. CPFH pulse sequence is the Freeman–Hill modification of the Carr–Purcell sequence [65, 71]. It employs a (0°, 180°) alternation of the π pulse phases. Additional future applications can be considered from the review of NMR pulse sequences for studying different systems [73].

The CPMG experiments on C15m collagen fibers (HL=0.6) at varying temperature obtained the R_2 data to calculate an activation energy of water molecular motion. For instance, according to the Arrhenius dependence of LnR_2 *vs* the reciprocal temperature (the range between 298 K and 263 K), E_a was estimated as 7.41 kJ/mol [35]. ($Ln\ R_1(1/T\ K)$ data resulted in $E_a \approx 6.18$ kJ/mol). The activation energy E_a calculated from relaxation rates R_2 measurements was associated with dipole-dipole interactions. Also, this E_a value may be compared with $E_a = 11$ kJ/mol from R_2 data for hydrated collagen at HL=0.33 [53]. Renou *et al.* studied these collagens with HL ≤ 0.33 indicating some difference in E_a values between different cross-link states of collagen and different water content [53].

The short transverse relaxation time (*e.g.* T_2 = 1.78 ms in C15m at HL = 0.6) could be attributed to water fraction in microfibrils, and slow T_2 component (T_2 =7.3 ms in C15m at HL = 0.6) was supposed to be associated with water molecules in the interfibrillar space. The fractions of water molecules associated with microfibrils are disturbed by influence of structural fraction as it was estimated because of near location to structural water of tropocollagen [35]. The second water fraction has water molecules with higher mobility than microfibrillar water fraction, and this fraction is influenced by addition of free water at increasing hydration level in collagen samples. The population of slow T_2 component increased in the collagens with HL = 0.87 in comparison with HL = 0.45 (and HL = 0.6). However, relative population of short T_2 component was always bigger than population of the long one. At the same time, the change in T_2 relaxation time of long component with increasing HL (T_2 increased in C15m fibers from 6.3 ms at HL = 0.45 to 10.3 ms at HL = 0.87) indicates that the thickness of water interfibrillar space changes. This is associated with the swelling of collagen matrix as the water content increases.

CONCLUDING REMARKS

The data of NMR relaxation methods in studying interaction of water with biomacromolecules in natural silk, nonoriented collagen fibers and oriented collagens with various cross-linking level have been considered in the chapter. The material analysed in the chapter highlighted the characteristics of water-biopolymers interactions based on the NMR data. The details can be taken into account for the proper implementation of NMR relaxation measurements. This is essential for understanding NMR relaxation methods and the processes of hydration / dehydration of natural biomaterials.

The NMR relaxation times (T_1 and T_2) have been studied in the collagen fibers oriented along the static magnetic field B_0. The data of NMR relaxation experiments on randomly oriented fibers of collagens with varying hydration level have been considered also. Cross-relaxation effect has been studied on silk and collagen samples with low water content. Correlation times as characteristics of molecular motions have been discussed to quantitatively compare the mobility of water and biopolymer macromolecules in natural silk fibers and collagen tissues with different level of cross-links and binding water molecules.

REFERENCES

[1] G.J.G. Davies, D.P. Knight, and F. Vollrath, "Chitin in the silk gland ducts of the spider *Nephila edulis* and the silkworm *Bombyx mori*", *PLoS One,* vol. 8, no. 8, p. e73225, 2013. [http://dx.doi.org/10.1371/journal.pone.0073225] [PMID: 24015298]

[2] Os. Hakimi, D.P. Knight, F. Vollrath, and P. Vadgama, "Spider and mulberry silkworm silks as compatible biomaterials", *Compos., Part B Eng.,* vol. 38, pp. 324-327, 2007. [http://dx.doi.org/10.1016/j.compositesb.2006.06.012]

[3] V.V. Rodin, and D.P. Knight, "Water self-diffusion in natural silk fibers as determined by the pulsed magnetic field gradient method", *Biophysics,* vol. 48, no. 3, pp. 404-410, 2003. [PMID: 12815853]

[4] V.V. Rodin, G.M. Reznichenko, and E.L. Vasina, "Properties of natural polymer fibers with low water content", *Mater. Sci.,* no. 1, pp. 34-42, 2004.

[5] V.V. Rodin, and D.P. Knight, "Natural materials by NMR data: cross-relaxation in macromolecules of natural silk", *Mater. Sci.,* no. 10, pp. 16-21, 2003.

[6] V.V. Rodin, "Methods of Magnetic Resonance in Studying Natural Biomaterials", In: *Encyclopedia of Physical Organic Chemistry,* Zerong Wang, Ed., Chapter 53. vol. 4, part 4. John Wiley & Sons, Inc., 2017, pp. 2861-2908.

[7] V.V. Rodin, G.M. Reznichenko, and E.L. Vasina, "Properties of natural silk fibers: deformation studies and NMR data", *Biophysics,* vol. 49, no. 6, pp. 918-926, 2004. [PMID: 15612542]

[8] V.V. Rodin, "NMR Relaxation in investigation of water in hydrated collagen of Lesser Spotted Dogfish", *Biophysics,* vol. 50, no. 2, pp. 219-225, 2005.

[9] V.V. Rodin, and D.P. Knight, "Molecular mobility in natural polymers: *Bombyx mori* silk with low water content as studied by ^{1}H DQF NMR", *Biophysics,* vol. 49, no. 5, pp. 730-737, 2004. [PMID: 15526463]

[10] S. Han, S.J. Gemmell, K.G. Helmer, P. Grigg, J.W. Wellen, A.H. Hoffman, and C.H. Sotak, "Changes in ADC caused by tensile loading of rabbit achilles tendon: evidence for water transport", *J. Magn. Reson.,* vol. 144, no. 2, pp. 217-227, 2000.

[http://dx.doi.org/10.1006/jmre.2000.2075] [PMID: 10828190]

[11] S.F. Othman, E.T. Curtis, S.A. Plautz, A.K. Pannier, S.D. Butler, and H. Xu, "MR elastography monitoring of tissue-engineered constructs", *NMR Biomed.,* vol. 25, no. 3, pp. 452-463, 2012. [http://dx.doi.org/10.1002/nbm.1663] [PMID: 21387443]

[12] G. Navon, U. Eliav, D.E. Demco, and B. Blümich, "Study of order and dynamic processes in tendon by NMR and MRI", *J. Magn. Reson. Imaging,* vol. 25, no. 2, pp. 362-380, 2007. [http://dx.doi.org/10.1002/jmri.20856] [PMID: 17260401]

[13] S. Flibotte, R.S. Menon, and A.L. MacKay, "Proton magnetic resonance of western red cedar", *Wood Fiber Sci.,* vol. 22, no. 4, pp. 362-376, 1990.

[14] U. Eliav, M. Komlosh, P.J. Basser, and G. Navon, "Characterization and mapping of dipolar interactions within macromolecules in tissues using a combination of DQF, MT and UTE MRI", *NMR Biomed.,* vol. 25, no. 10, pp. 1152-1159, 2012. [http://dx.doi.org/10.1002/nbm.2784] [PMID: 22362643]

[15] J. Wellen, K.G. Helmer, P. Grigg, and C.H. Sotak, "Spatial characterization of T_1 and T_2 relaxation times and the water apparent diffusion coefficient in rabbit Achilles tendon subjected to tensile loading", *Magn. Reson. Med.,* vol. 53, no. 3, pp. 535-544, 2005. [http://dx.doi.org/10.1002/mrm.20361] [PMID: 15723394]

[16] V.V. Rodin, *Magnetic Resonance Methods*. Press MIPhT: Moscow, 2004.

[17] A.L. MacKay, M. Bloom, M. Tepfer, and I.E. Taylor, "Broadline proton magnetic resonance study of cellulose, pectin and bean cell walls", *Biopolymers,* vol. 21, no. 8, pp. 1521-1534, 1982. [http://dx.doi.org/10.1002/bip.360210805]

[18] M. Estilaei, A. MacKay, C. Roberts, and J. Mayo, " ^{1}H NMR measurements of wet/dry ratio and T_1, T_2 distributions in lung", *J. Magn. Reson.,* vol. 124, no. 2, pp. 410-419, 1997. [http://dx.doi.org/10.1006/jmre.1996.1044] [PMID: 9169222]

[19] P.A. Jansson, *Deconvolution with Application in Spectroscopy.* Academic Press, 1984.

[20] Y. Takahashi, M. Gehoh, and K. Yuzuriha, "Structure refinement and diffuse streak scattering of silk (*Bombyx mori*)", *Int. J. Biol. Macromol.,* vol. 24, no. 2-3, pp. 127-138, 1999. [http://dx.doi.org/10.1016/S0141-8130(98)00080-4] [PMID: 10342756]

[21] K. Yamaguchi, Y. Kikuchi, T. Takagi, A. Kikuchi, F. Oyama, K. Shimura, and S. Mizuno, "Primary structure of the silk fibroin light chain determined by cDNA sequencing and peptide analysis", *J. Mol. Biol.,* vol. 210, no. 1, pp. 127-139, 1989. [http://dx.doi.org/10.1016/0022-2836(89)90295-7] [PMID: 2585514]

[22] D. Ebrahimi, O. Tokareva, N.G. Rim, J.Y. Wong, D.L. Kaplan, and M.J. Buehler, "Silk - its mysteries, how it is made, and how it is used", *ACS Biomater Sci Eng,* vol. 1, no. 10, pp. 864-876, 2015. [http://dx.doi.org/10.1021/acsbiomaterials.5b00152] [PMID: 27398402]

[23] E.M. Haacke, R.W. Brown, M.R. Thompson, and R. Venkatesan, *Magnetic resonance imaging: physical principles and sequence design.* Wiley-Liss: New York, 1999.

[24] W.E. Steinmetz, and C.R. Maher, "Magnetic resonance imaging on an NMR spectrometer. An experiment for the physical chemistry laboratory", *Concepts Magn. Reson. Part A,* vol. 30A, no. 3, pp. 133-139, 2007.

[25] P.Y. Ghi, D.J. Hill, and A.K. Whittaker, "NMR imaging of water sorption into poly(hydroxyethyl methacrylate-co-tetrahydrofurfuryl methacrylate)", *Biomacromolecules,* vol. 2, no. 2, pp. 504-510, 2001. [http://dx.doi.org/10.1021/bm000146q] [PMID: 11749213]

[26] V.V. Rodin, T. Cosgrove, and M. Murrey, "Magnetic resonance imaging and pulsed field gradient NMR in study of structure of polymer solutions and films", *Proceedings of the 6th Colloquium on Mobile Magnetic Resonance,* 2006, pp. 32-33 Aachen, Germany

[27] I. Jäger, and P. Fratzl, "Mineralized collagen fibrils: a mechanical model with a staggered arrangement of mineral particles", *Biophys. J.,* vol. 79, no. 4, pp. 1737-1746, 2000. [http://dx.doi.org/10.1016/S0006-3495(00)76426-5] [PMID: 11023882]

[28] P. Fratzl, K. Misof, I. Zizak, G. Rapp, H. Amenitsch, and S. Bernstorff, "Fibrillar structure and mechanical properties of collagen", *J. Struct. Biol.,* vol. 122, no. 1-2, pp. 119-122, 1998. [http://dx.doi.org/10.1006/jsbi.1998.3966] [PMID: 9724612]

[29] L.W. Jelinski, A. Blye, O. Liivak, C. Michal, G. LaVerde, A. Seidel, N. Shah, and Z. Yang, "Orientation, structure, wet-spinning, and molecular basis for supercontraction of spider dragline silk", *Int. J. Biol. Macromol.,* vol. 24, no. 2-3, pp. 197-201, 1999. [http://dx.doi.org/10.1016/S0141-8130(98)00085-3] [PMID: 10342765]

[30] L.S. Loo, R.E. Cohen, and K.K. Gleason, "Deuterium nuclear magnetic resonance of deuterium oxide in nylon 6 under active uniaxial deformation", *Polymer,* vol. 41, no. 21, pp. 7699-7704, 2000. [http://dx.doi.org/10.1016/S0032-3861(00)00118-X]

[31] Y. Takahashi, "Structural disorder of poly(vinylidene fluoride) form I: glides", *Macromolecules,* vol. 26, pp. 1471-1473, 1993. [http://dx.doi.org/10.1021/ma00058a044]

[32] V.S. Derinovskii, Y.R. Yarda, I.N. Zakirov, and M.P. Myasnikova, "Molecular motion and nuclear magnetic relaxation in filled poly(vinyl acetate)", *Polymer Sci. USSR,* vol. 32, no. 4, pp. 776-782, 1990. [http://dx.doi.org/10.1016/0032-3950(90)90403-S]

[33] E.R. Andrew, and R.G. Eades, "Separation of the intramolecular and intermolecular contributions to the second moment of the nuclear magnetic resonance spectrum", *Proc. Phys. Soc. Letters to the Editor,* vol. A66, no. 4, p. 415, 1953.

[34] F. Bashirov, and N. Gaisin, "Using the intramolecular contribution to the second moment of NMR line shape to detect site symmetry breakdown in molecular crystals", *Intern. J. Spectrosc,* vol. 1-2, pp. 1-6, 2015.

[35] V.V. Rodin, and V.A. Nikerov, "NMR relaxation and PFG NMR studies of water dynamics in oriented collagen fibres with different degree of cross-linking", *Curr. Tissue Eng.,* vol. 3, no. 1, pp. 47-61, 2014. [http://dx.doi.org/10.2174/2211542003666140626211652]

[36] V.V. Rodin, L. Foucat, and J.P. Renou, "Natural polymers according to NMR data: cross-relaxation in collagen macromolecules from two samples of connective tissues", *Biophysics,* vol. 49, no. 4, pp. 563-571, 2004. [PMID: 15458243]

[37] M-C. Vackier, B.P. Hills, and D.N. Rutledge, "An NMR relaxation study of the state of water in gelatin gels", *J. Magn. Reson.,* vol. 138, no. 1, pp. 36-42, 1999. [http://dx.doi.org/10.1006/jmre.1999.1730] [PMID: 10329223]

[38] S. Peto, P. Gillis, and V.P. Henri, "Structure and dynamics of water in tendon from NMR relaxation measurements", *Biophys. J.,* vol. 57, no. 1, pp. 71-84, 1990. [http://dx.doi.org/10.1016/S0006-3495(90)82508-X] [PMID: 2297563]

[39] R. Fechete, D.E. Demco, U. Eliav, B. Blümich, and G. Navon, "Self-diffusion anisotropy of water in sheep Achilles tendon", *NMR Biomed.,* vol. 18, no. 8, pp. 577-586, 2005. [http://dx.doi.org/10.1002/nbm.995] [PMID: 16320373]

[40] R. Blinc, V. Rutar, and I. Zupancic, "Proton NMR relaxation of adsorbed water in gelatin and collagen", *Appl. Magn. Reson.,* vol. 9, no. 2, pp. 193-216, 1995. [http://dx.doi.org/10.1007/BF03162042]

[41] U. Eliav, and G. Navon, "A study of dipolar interactions and dynamic processes of water molecules in tendon by ^{1}H and ^{2}H homonuclear and heteronuclear multiple-quantum-filtered NMR spectroscopy",

J. Magn. Reson., vol. 137, no. 2, pp. 295-310, 1999. [http://dx.doi.org/10.1006/jmre.1998.1681] [PMID: 10089163]

[42] G. Navon, U. Eliav, D.E. Demco, and B. Blümich, "Study of order and dynamic processes in tendon by NMR and MRI", *J. Magn. Reson. Imaging,* vol. 25, no. 2, pp. 362-380, 2007. [http://dx.doi.org/10.1002/jmri.20856] [PMID: 17260401]

[43] V.V. Rodin, L. Foucat, and J.P. Renou, "The dipolar interactions and dynamics of water molecules in collagen fibres from two connective tissues of different ages by ^{1}H double-quantum-filtered NMR spectroscopy", *Mater. Sci.,* no. 5, pp. 2-10, 2004.

[44] V.V. Rodin, "NMR approaches in studying dynamic molecules in porous materials", *J. Mater. Sci. Eng.,* vol. 3, no. 3, p. 245, 2014.

[45] Y. Seo, K. Ikoma, H. Takamiya, Y. Kusaka, L. Tsoref, U. Eliav, H. Shinar, and G. Navon, " ^{1}H double-quantum-filtered MR imaging as a new tool for assessment of healing of the ruptured Achilles tendon", *Magn. Reson. Med.,* vol. 42, no. 5, pp. 884-889, 1999. [http://dx.doi.org/10.1002/(SICI)1522-2594(199911)42:5<884::AID-MRM7>3.0.CO;2-0] [PMID: 10542346]

[46] L. Tsoref, H. Shinar, Y. Seo, U. Eliav, and G. Navon, "Proton double-quantum filtered MRI-a new method for imaging ordered tissues", *Magn. Reson. Med.,* vol. 40, no. 5, pp. 720-726, 1998. [http://dx.doi.org/10.1002/mrm.1910400512] [PMID: 9797155]

[47] V.V. Rodin, B.V. Sakharov, and D.P. Knight, "Hydrated polymer matrices: NMR relaxation and pulsed gradient in molecular mobility studies", *Plast. Massy,* no. 2, pp. 10-16, 2000.

[48] A. Kalk, and H.J.C. Berendsen, "Proton magnetic relaxation and spin diffusion in proteins", *J. Magn. Reson.,* vol. 24, no. 3, pp. 343-366, 1976.

[49] V.V. Rodin, B.V. Sakharov, V.N. Gerasimov, and D.P. Knight, "NMR relaxation in natural polymers (collagen spicules)", *Intern. Polym. Sci.Technol,* vol. 28, no. 3, pp. 70-78, 2001.

[50] A. Rochdi, L. Foucat, and J.P. Renou, "Effect of thermal denaturation on water-collagen interactions: NMR relaxation and differential scanning calorimetry analysis", *Biopolymers,* vol. 50, no. 7, pp. 690-696, 1999. [http://dx.doi.org/10.1002/(SICI)1097-0282(199912)50:7<690::AID-BIP2>3.0.CO;2-P] [PMID: 10547524]

[51] H.T. Edzes, and E.T. Samulski, "The measurement of cross-relaxation effects in the proton NMR spin-lattice relaxation of water in biological systems: hydrated collagen and muscle", *J. Magn. Reson.,* vol. 31, pp. 207-229, 1978.

[52] H.J.C. Berendsen, Specific interactions of water with biopolymers. *Water, A Comprehensive Treatise.,* F. Frank, Ed., vol. 5. Plenum Press: New York, 1975, pp. 293-330.

[53] J.P. Renou, M. Bonnet, G. Bielicki, A. Rochdi, and P. Gatellier, "NMR study of collagen-water interactions", *Biopolymers,* vol. 34, no. 12, pp. 1615-1626, 1994. [http://dx.doi.org/10.1002/bip.360341206] [PMID: 7849224]

[54] V.V. Rodin, B.V. Sakharov, V.N. Izmailova, and D.P. Knight, "Study of the collagen spicule structure by NMR relaxation and electron microscopy", *Russ. J. Biotechnol,* no. 4, pp. 257-265, 2001.

[55] A. Traore, L. Foucat, and J.P. Renou, " ^{1}H-NMR study of water dynamics in hydrated collagen: transverse relaxation-time and diffusion analysis", *Biopolymers,* vol. 53, no. 6, pp. 476-483, 2000. [http://dx.doi.org/10.1002/(SICI)1097-0282(200005)53:6<476::AID-BIP4>3.0.CO;2-8] [PMID: 10775063]

[56] J.Y. Exposito, U. Valcourt, C. Cluzel, and C. Lethias, "The fibrillar collagen family", *Int. J. Mol. Sci.,* vol. 11, no. 2, pp. 407-426, 2010. [http://dx.doi.org/10.3390/ijms11020407] [PMID: 20386646]

[57] A. Rochdi, L. Foucat, and J.P. Renou, "NMR and DSC studies during thermal denaturation of

collagen", *Food Chem.*, vol. 69, no. 3, pp. 295-299, 2000.
[http://dx.doi.org/10.1016/S0308-8146(99)00267-8]

[58] C.A. Miles, N.C. Avery, V.V. Rodin, and A.J. Bailey, "The increase in denaturation temperature following cross-linking of collagen is caused by dehydration of the fibres", *J. Mol. Biol.*, vol. 346, no. 2, pp. 551-556, 2005.
[http://dx.doi.org/10.1016/j.jmb.2004.12.001] [PMID: 15670603]

[59] R. Knauss, G. Fleischer, W. Gründer, J. Kärger, and A. Werner, "Pulsed field gradient NMR and nuclear magnetic relaxation studies of water mobility in hydrated collagen II", *Magn. Reson. Med.*, vol. 36, no. 2, pp. 241-248, 1996.
[http://dx.doi.org/10.1002/mrm.1910360211] [PMID: 8843378]

[60] J.S. Noggle, and R.E. Shirmer, *The Nuclear Overhauser Effect: Chemical Application.* Academic Press: New York, 1971.

[61] H.S. Gutowsky, D.W. McCall, and C.P. Slichter, "Nuclear magnetic resonance multiplets in liquids", *J. Chem. Phys.*, vol. 21, no. 2, pp. 279-292, 1953.
[http://dx.doi.org/10.1063/1.1698874]

[62] R.M. Henkelman, G.J. Stanisz, J.K. Kim, and M.J. Bronskill, "Anisotropy of NMR properties of tissues", *Magn. Reson. Med.*, vol. 32, no. 5, pp. 592-601, 1994.
[http://dx.doi.org/10.1002/mrm.1910320508] [PMID: 7808260]

[63] D.K. Kim, T.L. Ceckler, V.C. Hascall, A. Calabro, and R.S. Balaban, "Analysis of water-macromolecule proton magnetization transfer in articular cartilage", *Magn. Reson. Med.*, vol. 29, no. 2, pp. 211-215, 1993.
[http://dx.doi.org/10.1002/mrm.1910290209] [PMID: 8429785]

[64] B.M. Fung, and T.W. McGaughy, "Cross-relaxation in hydrated collagen", *J. Magn. Reson.*, vol. 39, pp. 413-420, 1980.

[65] G.E. Santyr, R.M. Henkelman, and M.J. Bronskill, "Variation in measured transverse relaxation in tissue resulting from spin locking with the CPMG sequence", *J. Magn. Reson.*, vol. 79, pp. 28-44, 1988.

[66] M. Kotecha, D. Klatt, and R.L. Magin, "Monitoring cartilage tissue engineering using magnetic resonance spectroscopy, imaging, and elastography", *Tissue Eng. Part B Rev.*, vol. 19, no. 6, pp. 470-484, 2013.
[http://dx.doi.org/10.1089/ten.teb.2012.0755] [PMID: 23574498]

[67] C. Migchelsen, and H.J.C. Berendsen, "Proton exchange and molecular orientation of water in hydrated collagen fibers. An NMR study of H_2O and D_2O", *J. Chem. Phys.*, vol. 59, no. 1, pp. 296-305, 1973.
[http://dx.doi.org/10.1063/1.1679805]

[68] J.R. Grigera, and H.J.C. Berendsen, "The molecular details of collagen hydration", *Biopolymers,* vol. 18, no. 1, pp. 47-57, 1979.
[http://dx.doi.org/10.1002/bip.1979.360180106]

[69] H. Takamiya, Y. Kusaka, Y. Seo, M. Noguchi, K. Ikoma, T. Morimoto, and Y. Hirasawa, "Characteristics of proton NMR T_2 relaxation of water in the normal and regenerating tendon", *Jpn. J. Physiol.*, vol. 50, no. 6, pp. 569-576, 2000.
[http://dx.doi.org/10.2170/jjphysiol.50.569] [PMID: 11173552]

[70] P.S. Belton, "NMR studies of hydration in low water content biopolymer systems", *Magn. Reson. Chem.*, vol. 49, suppl. 1, pp. S127-S132, 2011.
[http://dx.doi.org/10.1002/mrc.2848] [PMID: 22290703]

[71] D.L. Tzou, S.M. Lee, and H.N. Yeung, "Temperature dependence and phase transition of proton relaxation of hydrated collagen in intact beef tendon specimens *via* cross-relaxation spectroscopy", *Magn. Reson. Med.*, vol. 37, no. 3, pp. 359-365, 1997.

[http://dx.doi.org/10.1002/mrm.1910370310] [PMID: 9055225]

[72] R. Freeman, and H.D.W. Hill, "High-resolution study of NMR spin echoes: J-spectra", *J. Chem. Phys.*, vol. 54, no. 1, pp. 301-313, 1971. [http://dx.doi.org/10.1063/1.1674608]

[73] J.A. Parkinson, "Modern NMR pulse sequences in pharmaceutical R&D", *eMagRes,* vol. 4, no. 1, pp. 69-82, 2015.

CHAPTER 3

NMR Diffusion Studies of Water in Natural Biopolymers

Abstract: The chapter presents the results of one-dimensional NMR methods in studying molecular anisotropy and microstructure of collagen fibers from two connective tissues with different cross-linking level (fibers from adult steer and young calf) at humidity level (HL) of 0.6 g water per g dry mass. The apparent diffusion coefficients (D_{app}) have been studied in oriented collagen fibers (along the static magnetic field B_0) for two experimental cases: (1) the gradient was applied along the static magnetic field B_0 and (2) the gradient was switched on perpendicular to the magnetic field B_0. The dependences of D_{app} on diffusion time discovered a restriction diffusion of water for both gradient directions. A model of equally spaced parallel planes with permeable barriers, has been used to estimate a restricted distance and permeability coefficient. For both types of collagen fibers (adult and young) anisotropic diffusion of water has been discovered. Moreover, self-diffusion of water in fibers of natural silk (*Bombyx mori*) with HL=0.18 g H_2O per g dry mass has been studied by pulsed field gradient NMR stimulated echo at various diffusion times Δ between 10 and 200 ms. The analysis showed that the decrease in D_{app} with the increasing Δ due to the restricted diffusion. The results obtained were compared with published data on restricted diffusion in natural macromolecular systems with low water content.

Keywords: Apparent diffusion coefficient D_{app}, Free induction decay (FID), Natural Biopolymers, Pulsed Field Gradient (PFG) NMR, Spin-echo (SE), Spin-lattice (longitudinal) relaxation time T_1, Spin-spin (transverse) relaxation time T_2, Stimulated echo (STE).

^{1}H NMR STUDY OF THE SELF-DIFFUSION OF WATER IN CROSS-LINKED COLLAGENS

In many NMR diffusion studies, the dependence of echo intensity on gradient factor is described by single exponential function with an apparent diffusion coefficient D_{app} [1 - 10]. For example, the data obtained on C15m and C8y collagen fibers with HL = 0.6 g H_2O per g dry mass (Chapters **1**, **2**) also discovered monoexponential dependence. The value of D_{app} could be calculated in accordance with eq. (6) (Chapter **1**) [6]. It was found that diffusivities of water

Victor V. Rodin

in C15m collagen fibers, *i.e.,* in the samples with low level of cross-links are differed from those in the fibers with higher level of cross-links (C8y collagen). For instance, at Δ = 10 ms for *Z*-gradient, $D_z = 0.755 \times 10^{-9}$ m^2/s, and for the gradient along the *X*-axis, $D_x = 0.455 \times 10^{-9}$ m^2/s in C8y collagen, whereas in C15m fibers, $D_z = 0.510 \times 10^{-9}$ m^2/s, and $D_x = 0.31 \times 10^{-9}$ m^2/s. D_{app} was depending on diffusion time Δ. It decreased with increasing Δ for both types of collagen [6]. Such behaviour of $D_{app} = f(\Delta)$ is often observed in heterogeneous systems, such as fibrous tissues or porous materials [3, 11 - 20]. Sometimes, such a dependence of D_{app} on diffusion time was associated with the variations of intrinsic susceptibility [7, 17]. In heterogeneous systems, susceptibility variations result in the internal field gradient *g*. This inner gradient should be considered as an addition to external gradient [3, 7, 13 - 17]. According to works [12, 13, 17], the internal gradients can be approximated by *Gaussian* distribution function of width g_0, *i.e.*, $f(g)=(4/\pi)^{1/2}(1/g_0)\cdot\exp\{-(g/g_0)^2\}$ [20]. Then, the apparent diffusion coefficient should decrease with increasing Δ [3, 4, 11, 12] following to the formula $D_{app} = D\,(1 - \gamma^2 g_0^{\,2} D\Delta^3/4)$ [20]. Here, *D* is the diffusion coefficient not disturbed by the magnetic susceptibility effects. However, the data on apparent diffusion coefficients of water in C8y and C15m collagens cannot be described by this expression [4, 6]. In order to clarify the effects, which are responsible for the D_{app} dependence in C15m and C8y collagen fibers, the diffusion experiments with the *Cotts* bipolar gradients pulse sequence [15] have been carried out [6]. The results obtained showed good accordance between the data sets of STE and those of the *Cotts* pulse sequences. The results testified that background gradients due to heterogeneity of the magnet susceptibility give usually less effect than might have been expected. This is also in agreement with published data [3, 14] that the inner field gradients are essentially of no big influence on the diffusion data in hydrated collagen. Thus, $D_{app} = f(\Delta)$ dependence has been considered as a restriction phenomenon of water diffusion by barriers of macromolecular arrangement [4, 6]. If the diffusion is restricted (*i.e*, there are diffusion barriers), the mean square displacement of the protons increases with diffusion time according to $\langle x^2\rangle \sim \Delta^k$ with $k < 1$ [12 - 14, 16, 17] whereas apparent diffusion coefficient depends on Δ as $D_{app} \sim \Delta^{k-1}$. At completely restricted diffusion, $k = 0$ and $\langle x^2\rangle$ does not increase with diffusion time, therefore $D_{app} \sim 1/\Delta$. For the case of completely restricted diffusion, the fitting experimental curve $D_{app} = f(\Delta)$ should be done at $k = 0$. It was found, *e.g.*, for C15m collagen and C8y collagen, that for *Z*-direction (along the static magnetic field B_0) this is very good approximation [4]. For the *X*-direction, the fitted results showed partly restricted diffusion with parameters $k = 0.528$ (C15m collagen) and $k = 0.420$ (C8y collagen) (Figs. **1**, **2**). It was found that in the *Z*-direction, *k* value is near zero and it is possible to consider that in this case diffusion is almost completely restricted. For the *X*-direction, a restricted diffusion may be considered according to the dependence of $D_{app} \sim \Delta^{k-1}$ with k=0.42–0.53.

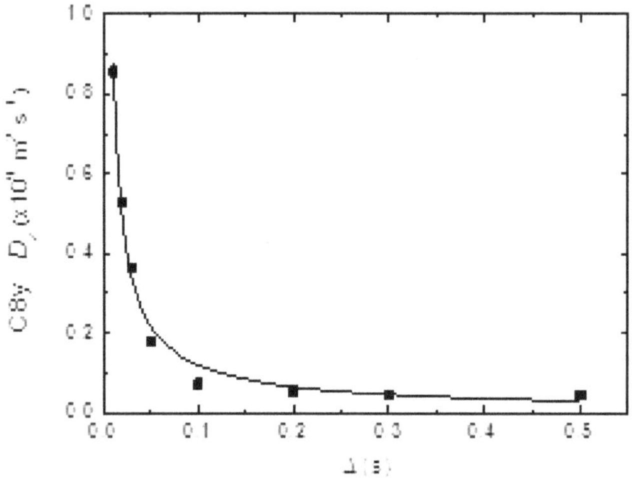

Fig. (1). Dependence of water diffusion coefficient D_{app} on the diffusion time for collagen fibers C8y oriented along static magnetic field B_0 (*Z*-direction). Gradient was applied along the *Z*-direction. The data presented have been obtained taking into account the cross-relaxation effect [6, 18, 19]. Solid line is the fit according to the model of equally spaced plane parallel permeable barriers developed by Tanner [1, 3, 11, 12] for NMR diffusion studies to estimate the sizes of restricted distance and permeability coefficient. *T*=298 K, and frequency is 400 MHz, HL=0.6.

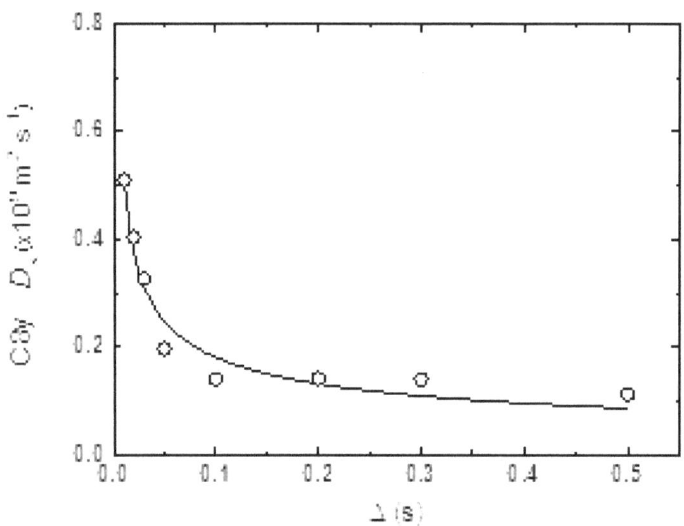

Fig. (2). Dependence of water diffusion coefficient D_{app} on the diffusion time for collagen fibers C8y oriented along static magnetic field B_0 (*Z*-direction). Gradient was applied along the *X*-direction. The data presented have been obtained taking into account the cross-relaxation effect [6, 18, 19]. Solid line is the fit according to the model of equally spaced plane parallel permeable barriers developed by Tanner [1, 3, 11, 12] for NMR diffusion studies to estimate the sizes of restricted distance and permeability coefficient. *T*=298 K, and frequency is 400 MHz, HL=0.6.

For short diffusion times, for instance, for Δ=10 ms, apparent diffusion coefficient D_{app} may be approximately considered as a free diffusion coefficient D_0 [3, 5, 7, 11, 12, 14]. Typical numerical results showing the theoretical variation of diffusion coefficient with diffusion time have been presented in [1 - 4]. When the diffusion time tends to zero, D_{app} approaches inter barrier diffusion coefficient D_0. At very long diffusion time Δ, it is numerically verified that D_{app} approaches value of D_{asym}, which may be calculated from $1/D_{asym} = 1/D_0 + 1/ap$ [1, 3, 4, 6], where a is the barrier spacing and p is the barrier permeability. A system of regularly spaced planar barriers with finite permeability p has been considered [1, 9 - 13]. The diffusion coefficient was calculated in the chosen system on the base of the model of water diffusion in fibrous tissue [7, 11 - 13, 20]. The diffusion data on collagens C15m and 8y were analysed and treated using the approach of Tanner [1, 3, 11]. According to Tanner *et al.*, the intermediate portions of the curves $D=f(\Delta)$ can be applied in calculation. In particular, Tanner used the locus of points, where D_{app} reaches its average value, $(D_0 + D_{asym})/2$. Fitting experimental data by the expression $D=f(t)$ based on the model for the restriction diffusion, it is possible to find $t_{1/2}$, corresponding to the average D_{app}.

The calculations for C8y and C15m collagens has been done using the approach of Tanner [11 - 13] with a connection between D and a as $D=a^2/12\Delta$ (when gradient is perpendicular to the barriers) to estimate the permeability constant p and restricted diffusion length a. These estimates for C15m collagen are a=7.39 μm, p=0.0009 cm s^{-1} (X-direction) and a=7.86 μm, p=0.0005 cm s^{-1} (Z-direction), whereas for C8y, a=8.18 μm, p=0.0011 cm s^{-1} (X-direction) and a=9.89 μm, p=0.00034 cm s^{-1} (Z-direction) [6]. All measured T_2 components characterising the water fractions in microfibril and interfibrillar spaces in C15m and C8y collagens contribute to the diffusion signal during all the diffusion experiments. Therefore, it is possible to consider the calculated values a=7.4–9.9 μm as characteristic sizes of spaces in collagen fibers with different cross-linking level.

Apparent diffusion coefficient D_{app} measured at small values Δ=10–20 ms was often considered in literature on NMR diffusion studies of collagen tissues as a free diffusion coefficient D_0. For instance, Ref. [21] considered the data on cartilage (HL=0.7–0.8) and used D_{app} at Δ=20 ms as D_0 to compare with diffusion coefficient D_w for water (D_w =2.3 × 10^{-9} m^2 s^{-1} at room temperature [4, 6, 9, 24]) and to obtain D_0/D_w =0.56–0.62. The comparison of this ratio D_0/D_w for C15m and C8y collagen fibers (*i.e.,* for adult and young samples, respectively) with HL=0.6 showed the higher diffusion coefficients in C8y samples for both directions of applied gradient (Z, X). For example, in the X-direction, D_{adult}/D_w = 0.22 and D_{young}/D_w = 0.15. In the case of gradient along the Z-direction, D_{adult}/D_w = 0.37 and D_{young}/D_w =0.25. When STE pulse sequence is applied in PFG studies, the results can be affected by CR phenomenon [4, 19]. The studies of C8y and C15m collagens showed that a cross-relaxation had a place in both oriented and random collagen fibers (Chapter **1**, Fig. **6**). The CR is responsible for the appearance of two T_1 components [6, 18, 19]. Spin-lattice relaxation has a place during time interval between the second and the third RF pulses in STE pulse sequence and respectively affects the echo-attenuation (Fig. **7**, *top*, Chapter **1**). This effect has been taken into account when the data $D=f(\Delta)$ are presented in Fig. (**1** and **2**).

As it was shown for the case of applied gradients along the *Z*-direction for C8y and C15m collagens, CR effect may be considered as very weak because the STE data treated according to eq. (6) of Chapter 1 without CR looked similar to the results treated with CR factor, in particular, for long diffusion times [4, 6]. However, in the case of applying gradient along *X*-direction, *D* values obtained at diffusion times Δ>100 ms and the fits with CR factor exceeded diffusion constants from the calculation according to eq. (6) of the Chapter 1 without CR factor (Fig. **3**). Therefore, permeability and restriction sizes were recalculated taking into account the CR factor [6].

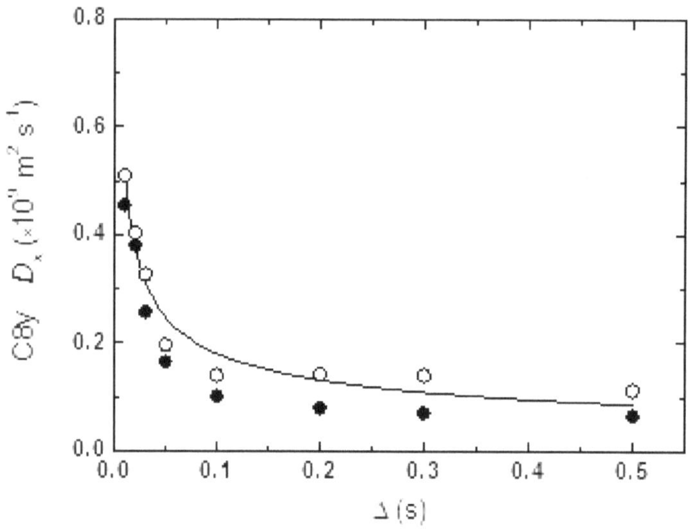

Fig. (3). Dependence of the water diffusion coefficient D_{app} on the diffusion time for collagen fibers C8y oriented along static magnetic field B_0 (*Z*-direction). HL = 0.6 g H_2O per g dried mass. *T*=298 K, and frequency is 400 MHz . The gradients were applied along the *X*-direction. Different data sets are presented: *filled circles* are the diffusion coefficients obtained with fitting STE attenuation curves by eq. (6), Chapter 1. *open circles* are the diffusion coefficients recalculated with cross-relaxation factor according to Refs. [6, 19]. Solid line is the same fitting as in the Fig. (**2**).

The studies of diffusivity of water in C8y and C15m collagens at HL=0.6 and Δ=10 ms showed that the ratio of D_z/D_x was equal to 1.66 (C15m collagen) and

1.678 (C8y collagen). However, for both collagens at Δ=20 ms this ratio was equal to 1.14. It is possible to consider this ratio $A_{z/x}=D_z/D_x$ as an indicator of anisotropy of diffusion properties and to compare measured $A_{z/x}$ with literature data [3, 5, 21]. For cartilage with HL=0.73–0.86, the ratio of diffusivities along two directions was about 1.06 [21]. The effect anisotropy in hydrated collagens (HL=1.7) was about 1.1 [22]. However, in the work of Henkelman *et al.*, tendon and cartilage did not show diffusion anisotropy [5]. In this work, the diffusion gradient remained oriented along the magnetic field B_0 and the sample was rotated. However, in the studies of hydrated collagens [6, 22], gradient could be applied in parallel or perpendicularly to the B_0 independently. Therefore, it was not necessary to rotate the samples and restart tuning and arrangement of experimental parameters separately. Diffusion anisotropy has been studied in the C15m and C8y collagens at small hydration level of the fibers (HL=0.6). In other studies [3, 5, 22], hydration level of collagens was always higher. For instance, HL=1.5 in the studies of Ref. [5].

In the study of the hydrated collagen in [22], a weak anisotropy effect was observed. The signal responsible for the amplitude of echo was compiled from different fractions of water. However, a behaviour of echo was explained by the effect of the main fraction, which is free water with the D_0 =2.15 × 10^{-9} m^2 s^{-1}. Thus, this main fraction results in isotropic mobility in total registered signal. The experiment has been carried out with the echo time 208 ms. Thus, only mobile water fractions with relatively long T_2 values defined echo-attenuation and diffusivity of molecules [22]. In the C15m and C8y collagens at HL=0.6, there was no free water and the echo signal was ruled by bound water protons [6].

The total proton signal is considered as a result of exchange between the two species (isotropic and anisotropic) sites [5, 21, 23 - 26]. When the value of HL is small, the averaged interaction remains in the fiber direction. According to the models of water–collagen interactions [3, 27 - 29], water molecules could have H bonds with solid protein base, both N–H and C=O groups, which are disposed out almost perpendicular to the fiber axis. As a result of possibility to have such water bonds, this type of anisotropy is explained by the arrangement of biopolymer chains and has been found only in collagen at small value of HL [22, 25, 28, 30].

SELF-DIFFUSION OF WATER IN *BOMBYX MORI* SILK AS STUDIED BY NMR

Natural silk *B. mori* combines many important properties: such as high strength, elasticity, sustainability to the action of the factors of environment (temperature, humidity, pH) [4]. In order to understand these properties better it is necessary to clarify the diffusion characteristics of water in silk fibers.

PFG NMR method was used in studying self-diffusion of water in different natural biopolymers and biomaterials for many years [1 - 4, 10, 13]. It has been shown that the diffusion of water in biomaterials could be often described by single diffusion coefficient, and its dependence on the diffusion time makes it possible to evaluate the structural properties of the hydrated natural materials [3, 4, 6, 17, 21, 22]. However, the use of PFG NMR for studying water diffusivity in silk fibers has been done first time only in the work [31]. To the moment, many details of water diffusivity in natural silk are not clarified. The section considers further how PFG NMR method in one direction is applying in the study of water self-diffusion in the natural silk fibers (*Bombyx mori*) with low hydration level. It has been found that the diffusion of water molecules in the silk fibroin fibers is restricted [4, 31]. Within a model of plane-parallel permeable barriers, middle restriction sizes for water diffusion and permeability coefficient were estimated. These results provide additional information on the structural properties and dynamics of molecules in natural materials.

Free induction decay after a 90° pulse (with a duration of 5-6 ms) in the hydrated silk was associated only with water protons, because of all measurements were made with a delay of 100 μs after the pulse. As a result, the FID signal due to the actual protons of macromolecules subsided very quickly because of very fast spin-spin relaxation time (T_2) and did not contribute to the total intensity of the measured magnetization signal of water. After FT of the FID into frequency-domain, the recorded spectrum was single resonance line. The registered signal was fitted by a Voight function (*e.g.,* eq. (3) from Chapter 2) which contains parameters ω_L and ω_G as *Lorentz* and *Gauss* widths of the NMR spectra at half of the peak height in frequency scale. The silk samples with water content HL = 0.18 had ω_L = 613±30 Hz and ω_G = 895±40 Hz. During silk fiber dehydration, the proton NMR signals broadened, and the shape lines were still described by the Voigt function [36]. For example, for dried fibers (HL = 0.065 g H_2O per g dry mass), the Voigt function parameters were ω_L = 1213±64 Hz and ω_G = 1290±70 Hz [4, 31, 37].

The apparent diffusion coefficients D_{app}, which were obtained from the slope of echo-attenuation curves in STE experiments, are presented on the Fig. (**4**). The values of D_{app} decreased with the increase of diffusion time Δ. For molecules diffusing in a porous medium or in medium of planar barriers with distance *a* between the ones different timescales can be distinguished [1, 7 - 9, 14, 16, 32]. For instance, at small Δ values, $(\langle z^2 \rangle)^{1/2} \ll a$, and few molecules are influenced by the restrictions. Thus, it is possible to consider that $D_{\Delta=10ms} \approx D_0$, where D_0 is the bulk diffusion coefficient in the absence of barriers [3, 6, 7, 17, 20]. At intermediate Δ values, $(\langle z^2 \rangle)^{1/2} \sim a$, and an apparent diffusion coefficient as a function of Δ is decreasing with increasing Δ due to the increasing number of molecules that reach the barriers. With barriers presented during diffusion time, $<z^2>$ becomes less than that estimated from *Gaussian* diffusion [1, 12, 38]. Therefore, the value of apparent diffusion coefficient D_{app} will decrease. In the case of permeable walls or connections between them, the diffusion coefficient reaches a constant value D_{asym} reflecting the long range permeability [1, 17]. For nonpermeable barriers or for barriers with low permeability, D_{asym} value tends to zero [1, 22, 33]. According to the approaches [1, 17, 31 - 34, 38], from the analysis of such data one can, in principle, obtain information on the geometry of the domains in which restricted diffusion has a place. In the *B.mori* silk fibers with HL=0.18, the values D_{app} decreased with increasing Δ from D_0=0.146×10^{-9} m^2·s^{-1} to the values $D_{\Delta=200ms}$=0.008×10^{-9} m^2 s^{-1}. The D_{app} values were lower than those observed in pure water (D_w=2.3×10^{-9} m^2·s^{-1}) in circa 16 times (at Δ=10 ms). This demonstrated thereby that the water in the natural silk with HL = 0.18 has a very low molecular mobility as compared with the free liquid water.

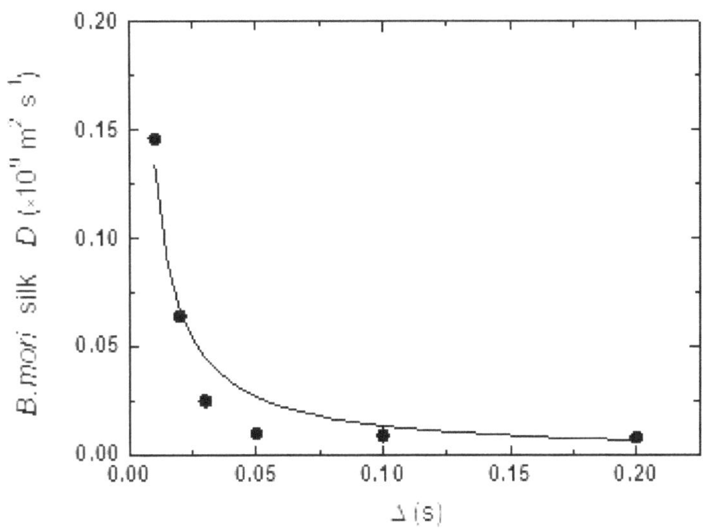

Fig. (4). Dependence of the water diffusion coefficient D_{app} on diffusion time Δ for randomly oriented fibers of natural silk *B. mori* at HL=0.18 g H_2O per g dried mass. T=298 K, and frequency is 400 MHz. Solid line is the result of fitting the experimental data in the model with restricted diffusion according to the dependence $D \sim \Delta^{k-1}$ (the details are in the text) at k=0.

Some factors resulting in mentioned above diffusion coefficients dependence on diffusion time (exchange, effect of magnetic susceptibility, restriction *etc.*) have been analyzed in Refs. [4, 31]. Fast exchange between different compartments is considered sometimes in the literature as possible factor affecting the apparent diffusion coefficients [7, 17]. However, an analysis of the echo-attenuation curves and a calculation of several diffusion constants are depending on special NMR approaches based on experimental T_2 components [3, 22, 35].

The results showed that the water fraction with a very low T_2 value (≈0.1÷0.2 ms or less) apparently does not affect the STE signal recorded in the investigated samples of natural silk. For silk samples with HL=0.065, the estimate of the spin-spin relaxation time T_2 has been done from the line width of ^{1}H NMR spectrum giving T_2=0.26 ms [31, 36, 37].

According to the analysis in Ref. [31], the internal field gradients, although present, but can not specify the observed self-diffusion of water in the natural silk. In comparison with this, Ref. [14] does not exclude the role of internal field gradients of magnetic field (at least in their contribution to the value of the measured relaxation time T_2) in hydrated collagen. In the publications on the self-diffusion of water in collagen [3, 14, 22], it has been shown that the internal gradients are not significant enough to have a noticeable effect on the self-diffusion of water and lead to the observed decrease in the value D_{app} with increasing diffusion time. Thus, it is considered that the restricted diffusion of water is responsible for the relationship $D_{app}=f(\Delta)$ in the natural silk *B.mori* [31]. This is due to barriers of restriction which are created by macromolecular chains of silk fibroin. The diffusion coefficient of water in natural silk depends on Δ according to $D_{app}\sim\Delta^{k-1}$. Thus, experimental data $D_{app}=f(\Delta)$ for *B. mori* silk are well approximated by a model function for restricted diffusion (k = 0) as $D_{app} = m/\Delta$ with adjustable parameter $m = 1.33\times10^{-3}$. This model curve is presented in Fig. (**4**). The diffusion coefficients of water in natural silk *B.mori* decrease from D_0 to asymptotic value D_{asym} with increasing Δ from 10 ms to 200 ms. An analysis of this decrease can be done presenting the data in the same coordinates, which were used in the literature, *i.e.*, through reduced diffusion time $T_r=D_0 t_d/a^2$ and D_{app} values normalized to inter-barrier diffusivity D_0 [1, 7, 22]. In reality, the value of D_{app} is depending on diffusion time, barrier permeability, and geometry/structure of the sample. It is possible to consider the value of the diffusion time t_d, which is associated with an average diffusivity $D(t_d) = (D_0 + D_{asym})/2$. For instance, Refs. [3, 7, 12, 33] considered the system with the finite barrier permeability p and the barrier spacing a. According to [7, 17, 20, 33], D_{app} reduced to the asymptotic self-diffusion coefficient D_{asym}, *i.e.*, the value which relates D_0 by the equation $D_{asym}=D_0/(1+D_0/ap)$ [1, 12].

Ref. [31] presented a typical dependence of the reduced diffusion coefficient $D(t)/D_0$ for hydrated natural silk as a function of the reduced diffusion time $T_r = D_0 t_d/a^2$. Therefore, the results for the natural silk *B.mori* can be compared with other published data considered within the model of plane-parallel barriers [1, 31]. A comparison of medium D_{app} values for some reduced permeabilities P (which has no dimension as the $P=ap/D_0$ [1]) shows that for more higher permeability P, a decay $D(t)/D_0$ has less sharp slope and reaches a constant value $D(t)/D_0=1$ for infinite permeability $P=\infty$. In the case of the use of the gradient, perpendicular to the plane of the barriers, a diffusivity D is related to diffusion time Δ and barrier spacing a as follows $D=a^2/12\cdot\Delta$) [11, 22, 31, 38]. This relation results in recalculation of the *B. mori* silk data in reduced diffusion coefficient and reduced permeability. Within the model of diffusion with permeable barriers, an average estimate of diffusion length resulted in $a\approx6.9$ μm, and reduced permeability $P\approx0.0575$. The permeability is the $p\approx2.1\times10^{-4}$ cm/s. This estimate of barrier distance is the averaged distance among all possible barrier characteristics of natural silk *B. mori* fibers [31].

Electronic microscopy (EM) data on natural silk [36, 37, 39] showed that a fiber diameter is estimated as 15-30 μm, and the fibers consist of 6-10 microfibrils. The dimensions of medium barrier distances obtained in PFG NMR study on *B. mori* silk, in general, were comparable to those of EM data. Thus, on the base of the data [31], it is possible to suppose that the boundaries of obstacles for diffusion of water molecules in *B. mori* natural silk could be associated with a surface of the fibers and microfibers.

As it was showed in [31 and the Refs. therein], the values of D_{app} are sensitive to the structural organization of the fibrillar polymers only on the scale exceeding 1 μm. This defines thereby that the limits of "structure resolution" of STE method in studying natural materials are caused by the maximum D_{app} value. The maximum measured $D_{app} = 0.146\times10^{-9}$ $m^2 \cdot s^{-1}$ corresponds to the diffusion distance 1.7 μm (at Δ=10 ms). In the study [31], the interval of very short diffusion times (<10 ms) remained unstudied. However, according to [1], it allows a possibility of the existence of such closely spaced barriers that changes in D_{app} occur on the intervals of shorter diffusion times than those used in the experiments (≥ 5-10 ms) [1, 31]. In order to take into account the barriers, which are not in the detected area, it is proposed [1] to consider the value of D_{app} at the shortest diffusion time (5-10 ms) as asymptotic diffusion constant (D_{asym}) and to present D_0 as the diffusion coefficient of bulk water. Following this approach in the short diffusion times range, it is possible to get for natural silk *B. mori* the relation $1+1/P = D_0/D_{asym} \leq 15$. Thus, this estimated reduced permeability range as $P\geq0.07$ [31]. For the barriers with such reduced permeability it results in the following experimental range for the reduced diffusion times $0.1<T_r<0.5$. This range can give a strong decrease in the dependence of $D(t)/D_0$ on time. In comparison with this, for the same range of diffusion times (for $T_r\leq0.3$), an assessment of the barriers [1] results in $a < 3$ μm. Such calculated barriers were not detected in STE experiment on the muscular tissue of the frog: the experimental dependency $D_{app} = f(\Delta)$ resulted in average restriction size as ~ 50 μm [1].

Probably, search of diffusion restrictions with barrier spacing circa ~1 μm or less in heterogeneous biological systems must still rely on the use of alternative pulse sequences for realisation of PFG NMR experiments in a short diffusion time limit.

CONCLUDING REMARKS

The chapter considered how restricted diffusion, an anisotropy of water motion, and microstructure of natural biopolymers could be studied by one-dimensional PFG NMR methods. Probably, the methods probed in the study on C15m and C8y collagens can be fruitful in studying other collagen containing tissues with different degree of binding water molecules and different number of cross-links between macromolecular chains.

Findings herein suggest that it is possible to register diffusion anisotropy in natural fibers by one-dimensional PFG NMR. These approaches are potentially applicable in the diagnosis and monitoring the collagen connective tissues at the diseases associated with macromolecular reorganization and defects in cross-links. The methods are suitable for investigating many kinds of porous medical biomaterials. The questions like how tissue engineering processes and modifications with crosslinks affect the diffusion anisotropy of water in collagen scaffolds could be in the area of these studies.

The study of *B. mori* natural silk clarified the details of self-diffusion of water in dependence on diffusion times. The considered findings discovered that the decisive role in the dependence $D_{app} = f(\Delta)$ can be attributed to the restricted diffusion of water molecules. The barrier spacing and the barrier permeability coefficient were evaluated. It is shown that the diffusion constants measured in weakly hydrated natural silk are significantly lower than those in aqueous solutions. These results complement and extend the information on the properties of water in heterogeneous natural biomaterials.

REFERENCES

[1] J.E. Tanner, "Self diffusion of water in frog muscle", *Biophys. J.*, vol. 28, no. 1, pp. 107-116, 1979. [http://dx.doi.org/10.1016/S0006-3495(79)85162-0] [PMID: 318065]

[2] V.V. Rodin, "NMR Relaxation in investigation of water in hydrated collagen of Lesser Spotted Dogfish", *Biophysics*, vol. 50, no. 2, pp. 219-225, 2005.

[3] A. Traore, L. Foucat, and J.P. Renou, " ^{1}H NMR study of water dynamics in hydrated collagen: transverse relaxation-time and diffusion analysis", *Biopolymers*, vol. 53, no. 6, pp. 476-483, 2000. [http://dx.doi.org/10.1002/(SICI)1097-0282(200005)53:6<476::AID-BIP4>3.0.CO;2-8] [PMID: 10775063]

[4] V.V. Rodin, "Methods of Magnetic Resonance in Studying Natural Biomaterials", In: *Encyclopedia of Physical Organic Chemistry*, Zerong Wang, Ed., Chapter 53. vol. 4, part 4. John Wiley & Sons, Inc., 2017, pp. 2861-2908. Available at: https://www.wiley.com/en-us/Encyclopedia+of+Physical+Organic+Chemistry%2C+6+Volume+Set-p-9781118470459

[5] R.M. Henkelman, G.J. Stanisz, J.K. Kim, and M.J. Bronskill, "Anisotropy of NMR properties of tissues", *Magn. Reson. Med.*, vol. 32, no. 5, pp. 592-601, 1994. [http://dx.doi.org/10.1002/mrm.1910320508] [PMID: 7808260]

[6] V.V. Rodin, and V.A. Nikerov, "NMR relaxation and PFG NMR studies of water dynamics in oriented collagen fibres with different degree of cross-linking", *Curr.Tissue Eng.*, vol. 3, no. 1, pp. 47-61, 2014. [http://dx.doi.org/10.2174/2211542003666140626211652]

[7] A. Szafer, J. Zhong, and J.C. Gore, "Theoretical model for water diffusion in tissues", *Magn. Reson. Med.*, vol. 33, no. 5, pp. 697-712, 1995. [http://dx.doi.org/10.1002/mrm.1910330516] [PMID: 7596275]

[8] J. Hrabe, G. Kaur, and D.N. Guilfoyle, "Principles and limitations of NMR diffusion measurements", *J. Med. Phys.*, vol. 32, no. 1, pp. 34-42, 2007. [http://dx.doi.org/10.4103/0971-6203.31148] [PMID: 21217917]

[9] V.V. Rodin, *Magnetic Resonance Methods*. Press MIPhT: Moscow, 2004.

[10] P.T. Callaghan, *Principles of Nuclear Magnetic Resonance Microscopy*. Clarendon Press: Oxford, 2001, p. 492.

[11] J.E. Tanner, and E.O. Stejskal, "Restricted self-diffusion of protons in colloidal systems by the pulsed-gradient, spin-echo method", *J. Chem. Phys.*, vol. 49, no. 4, pp. 1768-1777, 1968. [http://dx.doi.org/10.1063/1.1670306]

[12] J.E. Tanner, "Transient diffusion in system partitioned by permeable barriers. Application to NMR measurements with a pulsed field gradient", *J. Chem. Phys.*, vol. 69, no. 4, pp. 1748-1754, 1978. [http://dx.doi.org/10.1063/1.436751]

[13] J.E. Tanner, "Intracellular diffusion of water", *Arch. Biochem. Biophys.*, vol. 224, no. 2, pp. 416-428, 1983. [http://dx.doi.org/10.1016/0003-9861(83)90228-X] [PMID: 6347071]

[14] R. Knauss, G. Fleischer, W. Gründer, J. Kärger, and A. Werner, "Pulsed field gradient NMR and nuclear magnetic relaxation studies of water mobility in hydrated collagen II", *Magn. Reson. Med.*, vol. 36, no. 2, pp. 241-248, 1996. [http://dx.doi.org/10.1002/mrm.1910360211] [PMID: 8843378]

[15] R.M. Cotts, M.J.R Hoch, T. Sun, and J.T. Markert, "Pulsed field gradient stimulated echo methods for improved NMR diffusion measurements in heterogeneous systems", *J. Magn. Reson.*, vol. 83, no. 2, pp. 252-266, 1989.

[16] R. Trampel, J. Schiller, L. Naji, F. Stallmach, J. Kärger, and K. Arnold, "Self-diffusion of polymers in cartilage as studied by pulsed field gradient NMR", *Biophys. Chem.*, vol. 97, no. 2-3, pp. 251-260, 2002. [http://dx.doi.org/10.1016/S0301-4622(02)00078-9] [PMID: 12050014]

[17] R.P. Kennan, J.H. Zhong, and J.C. Gore, Effects of magnetic susceptibility variations on the apparent diffusion measured by NMR. In: *Diffusion and Perfusion: Magnetic Resonance Imaging.*, D. Le Bihan, Ed., Raven Press: New York, 1995, pp. 110-121.

[18] V.V. Rodin, L. Foucat, and J.P. Renou, "Natural polymers according to NMR data: cross-relaxation in collagen macromolecules from two samples of connective tissues", *Biophysics,* vol. 49, no. 4, pp. 563-571, 2004. [PMID: 15458243]

[19] L.J.C. Peschier, J.A. Bouwstra, J. de Bleyser, H.E. Junginger, and J.C. Leyte, "Cross-relaxation effects in pulsed-field-gradient stimulated-echo measurements on water in a macromolecular matrix", *J. Magn. Reson. B.*, vol. 110, no. 2, pp. 150-157, 1996. [http://dx.doi.org/10.1006/jmrb.1996.0024]

[20] M. Kveder, G. Lahajnar, R. Blinc, and I. Zupancic, "Non-Brownian water self-diffusion in lung tissue", *Magn. Reson. Med.*, vol. 6, no. 2, pp. 194-198, 1988. [http://dx.doi.org/10.1002/mrm.1910060207] [PMID: 3367776]

[21] D. Burstein, M.L. Gray, A.L. Hartman, R. Gipe, and B.D. Foy, "Diffusion of small solutes in cartilage

as measured by nuclear magnetic resonance (NMR) spectroscopy and imaging", *J. Orthop. Res.*, vol. 11, no. 4, pp. 465-478, 1993.
[http://dx.doi.org/10.1002/jor.1100110402] [PMID: 8340820]

[22] A. Traore, *Etude par RMM et diffusion des neutrons de la dynamique de l'eau dans le collagene nativ et dénaturé* These. Universite Blaise Pascal, 1997, p. 238.

[23] S. Peto, P. Gillis, and V.P. Henri, "Structure and dynamics of water in tendon from NMR relaxation measurements", *Biophys. J.*, vol. 57, no. 1, pp. 71-84, 1990.
[http://dx.doi.org/10.1016/S0006-3495(90)82508-X] [PMID: 2297563]

[24] H.J.C. Berendsen, Specific interactions of water with biopolymers. In: *Water, A Comprehensive Treatise.*, F. Frank, Ed., vol. 5. Plenum Press: New York, 1975, pp. 293-330.

[25] B.M. Fung, and T.W. McGaughy, "Cross-relaxation in hydrated collagen", *J. Magn. Reson.*, vol. 39, pp. 413-420, 1980.

[26] D.L. Tzou, S.M. Lee, and H.N. Yeung, "Temperature dependence and phase transition of proton relaxation of hydrated collagen in intact beef tendon specimens *via* cross-relaxation spectroscopy", *Magn. Reson. Med.*, vol. 37, no. 3, pp. 359-365, 1997.
[http://dx.doi.org/10.1002/mrm.1910370310] [PMID: 9055225]

[27] C.A. Miles, N.C. Avery, V.V. Rodin, and A.J. Bailey, "The increase in denaturation temperature following cross-linking of collagen is caused by dehydration of the fibres", *J. Mol. Biol.*, vol. 346, no. 2, pp. 551-556, 2005.
[http://dx.doi.org/10.1016/j.jmb.2004.12.001] [PMID: 15670603]

[28] C. Migchelsen, and H.J.C. Berendsen, "Proton exchange and molecular orientation of water in hydrated collagen fibers. An NMR study of H_2O and D_2O", *J. Chem. Phys.*, vol. 59, no. 1, pp. 296-305, 1973.
[http://dx.doi.org/10.1063/1.1679805]

[29] M. Granke, M.D. Does, and J.S. Nyman, "The role of water compartments in the material properties of cortical bone", *Calcif. Tissue Int.*, vol. 97, no. 3, pp. 292-307, 2015.
[http://dx.doi.org/10.1007/s00223-015-9977-5] [PMID: 25783011]

[30] J.R. Grigera, and H.J.C. Berendsen, "The molecular details of collagen hydration", *Biopolymers*, vol. 18, pp. 47-57, 1979.
[http://dx.doi.org/10.1002/bip.1979.360180106]

[31] V.V. Rodin, and D.P. Knight, "Water self-diffusion in natural silk fibers as determined by the pulsed magnetic field gradient method", *Biophysics*, vol. 48, no. 3, pp. 404-410, 2003.
[PMID: 12815853]

[32] G.H. Sorland, B. Hafskjold, and O. Herstad, "A stimulated-echo method for diffusion measurements in heterogeneous media using pulsed field gradients", *J. Magn. Reson.*, vol. 124, no. 1, pp. 172-176, 1997.
[http://dx.doi.org/10.1006/jmre.1996.1029]

[33] P. Stilbs, "Fourier transform pulsed-gradient spin-echo studies of molecular diffusion", *Progr. NMR Spectr*, vol. 19, no. 1, pp. 1-45, 1987.
[http://dx.doi.org/10.1016/0079-6565(87)80007-9]

[34] M.D. Silva, K.G. Helmer, J-H. Lee, S.S. Han, C.S. Springer Jr, and C.H. Sotak, "Deconvolution of compartmental water diffusion coefficients in yeast-cell suspensions using combined T_1 and diffusion measurements", *J. Magn. Reson.*, vol. 156, no. 1, pp. 52-63, 2002.
[http://dx.doi.org/10.1006/jmre.2002.2527] [PMID: 12081442]

[35] D. Van Dusschoten, P.A. De Jager, and H. Van As, "Extracting diffusion constants from echo-time-dependent PFG NMR data using relaxation-time information", *J. Magn. Reson. A*, vol. 116, no. 1, pp. 22-28, 1995.

[36] V.V. Rodin, G.M. Reznichenko, and E.L. Vasina, "Properties of natural silk fibers: deformation

studies and NMR data", *Biophysics,* vol. 49, no. 6, pp. 918-926, 2004. [PMID: 15612542]

[37] V.V. Rodin, G.M. Reznichenko, and E.L. Vasina, "Properties of natural polymer fibers with low water content", *Mater. Sci.,* no. 1, pp. 34-42, 2004.

[38] P.T. Callaghan, *Principles of Nuclear Magnetic Resonance Microscopy.* Clarendon Press: Oxford, 2001, p. 492.

[39] V.N. Gerasimov, "Problems of biological and ecological safety", *Proceedings of Intern. Conf.,* 2000, pp. 322-323, Obolensk, Russia

CHAPTER 4

Collagen Tissues with Different Degree of Cross-Links and Natural Silk as Studied by ^{1}H DQF NMR

Abstract: ^{1}H-^{1}H residual dipolar interactions in water molecules in oriented (along the static magnetic field B_0) and randomly oriented collagen fibers from two connective tissues with different cross-linking level (fibers from adult steer and young calf) at humidity level (HL) of 0.6 g water per g dry mass were investigated by ^{1}H double-quantum-filtered (DQF) NMR spectroscopy. It was observed that intensities of DQF signal in fibers oriented along outer static magnetic field exceeded significantly the DQF intensities in nonoriented fibers. The results were analysed on the basis of tensor description of DQF signal formation at the action of special sequence of radio frequency pulses with suitable phase cycling. The DQF signals dependencies upon creation time forming the ($T_{2,+1}$, $T_{2,-1}$) tensors of the second rank were compared for oriented and nonoriented collagen fibers with different cross-links. A role of exchange has been clarified in an additional study of temperature influence on DQF signals. The registered ^{1}H DQF NMR signals were due to the residual dipolar interactions (RDIs) between the protons of water molecules interacting with oriented collagen fibers. Molecular mobility of water in the fibers of *Bombyx mori* natural silk has been studied using the DQF NMR technique and single-pulse ^{1}H NMR. The observed ^{1}H DQF NMR signal in the natural silk fibers with low water content testified about local order and anisotropic motion of water molecules. DQF spectra in both collagens and silk fibers were analysed within the theory of ^{1}H DQF NMR on RDIs in the systems with anisotropic mobility. When a humidity level of natural silk fibers was increased to 0.18 g H_2O per g dry mass, no DQF signals and RDIs were detected.

Keywords: *Bombyx mori* natural silk, Collagen fibers C15m and C8y, Creation time, Double-quantum-filter (DQF) NMR, Evolution time, Free induction decay (FID), Humidity level (HL), Natural biopolymers, Residual dipolar interaction (RDI), Spin-lattice (longitudinal) relaxation time T_1, Spin-spin (transverse) relaxation time T_2.

^{1}H DQF NMR SPECTROSCOPY IN STUDYING COLLAGENS WITH DIFFERENT DEGREE OF CROSS-LINKS

The collagen fibers of the aponeurosis of 15 month old calf (C15m) and those of 8 year old steer (C8y) at humidity level (HL) of 0.6 g H_2O per g dry mass were the

Victor V. Rodin

subject of ^{1}H DQF NMR studies. Additional experimental details for these samples are in Chapter 1. The detection of ^{1}H DQF NMR signal was done using the pulse sequence: $90°_{\varphi}$–$(\tau_{creat}/2)$–$180°_{\varphi 1}$–$(\tau_{creat}/2)$–$90°_{\varphi}$–t_1–$90°_{\varphi 2}$–t_2 (acquisition) [1], which is shown in Fig. (**9**) of Chapter 1.

Early work presented the ^{1}H DQF NMR spectra of C15m- and C8y-oriented collagen fibers and showed that the amplitude of the DQF signal was dependent on the evolution time t_1. The intensity of DQF NMR signal decreased with increase in the evolution time from 100 μs to 2 ms (at fixed τ_{creat}) [1, 2] (Fig. **1**).

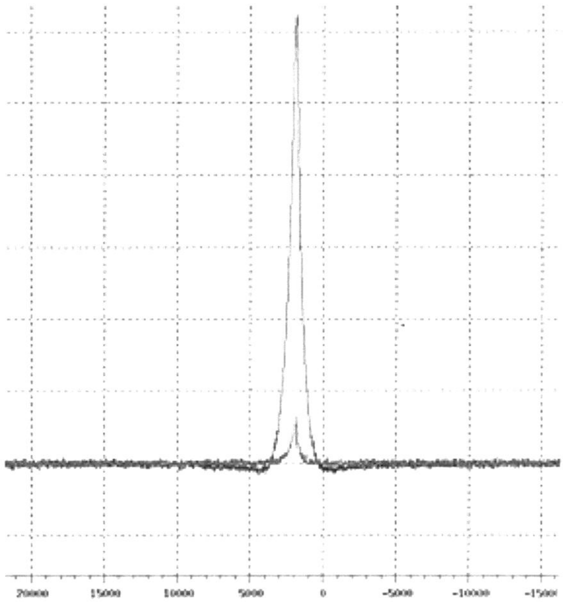

Fig. (1). ^{1}H DQF NMR spectra (an intensity is in arbitrary units) of oriented fibers of C15m collagen at varying evolution time from t_1=0.1 ms (high intensity) to t_1=2 ms (low intensity). T=298 K, and resonance frequency for protons is 400 MHz. Creation time is 0.1 ms, frequency scale is in Hz.

A comparative analysis of the ^{1}H DQF NMR spectra in C15m and C8y collagens at the same creation (100 μs) and evolution (100 μs) time has been done. Fig. (**2**) shows several DQF spectra for C15m oriented collagen fibers at various creation times. There was no signal of free water with isotropic motion in those spectra [1]. This indicated the efficiency of DQF sequence to study anisotropic motion as it was considered in other published works on collagen-containing tissues [3 - 11]. The results of fitting experimental DQF spectra as a combination of two antiphase *Lorentzians* (negative and positive peaks) with broad (ω_L ≈1500–2000 Hz) and narrow (ω_L ≈500–800 Hz) components are presented in Table **1**.

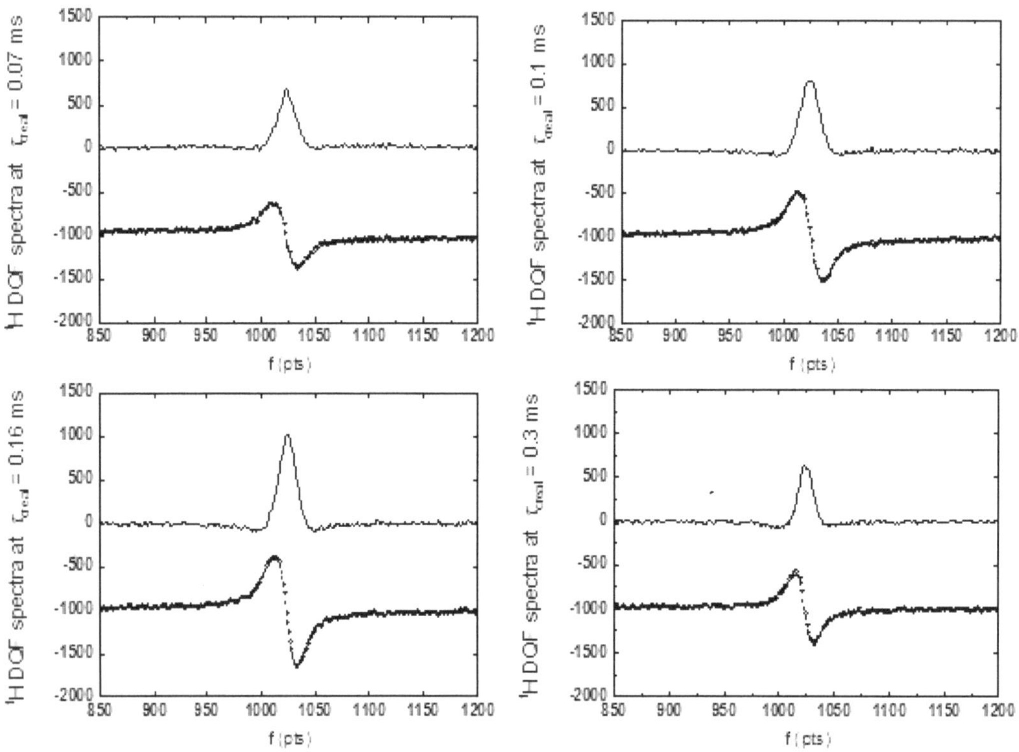

Fig. (2). ^{1}H DQF NMR spectra ("in phase" and "out of phase" modes) of C15m collagen fibers (oriented along the static magnetic field B_0) with increasing creation times from 0.07 ms (top, left) to 0.3 ms (bottom, right). The signal intensity increased towards spectrum of τ_{creat}=0.16 ms and then decreased again at τ_{creat}=0.3 ms. T=298 K, and resonance frequency for protons is 400 MHz. Experimentally obtained FIDs were transformed into frequency-domain using software *MestreC*. Frequency scale is presented in points: 50 pts = 6 kHz.

At a constant creation time, the amplitude A~$\exp(-s_{dq}\cdot t_1)$ in the expression of M_{xy} for the ^{1}H DQF NMR signal (Chapter 1) decayed exponentially with a rate constant of s_{dq}. The value of this rate constant s_{dq} for C8y collagen (~4.7 ms^{-1}) roughly estimated from experiments with creation time of 100 μs [1] was 5–6 times larger than that for C15m fibers (~0.9 ms^{-1}). Comparative DQF study has been carried out on the oriented fibers of C15m and C8y collagens and randomly distributed fibers [1, 2]. In the experiments with nonoriented fibers, one observed smaller DQF signal intensities (Fig. **3**).

In randomly distributed fibers, an angle θ_e (orientation of each particular fiber to main static field B_0) can accept different random values. Thus, the portion of fibers with $\theta = 0^\circ$ (at this angle, the fiber orientation is along the magnetic field B_0 resulting in maximal intensity of DQF signal) is significantly smaller than that in the sample where all fibers were oriented along the magnetic field B_0. This work

[1] studied the intensities of ^{1}H DQF NMR signals as a functions of spherical tensors ($T_{2,+1}$, $T_{2,-1}$)' creation times τ_{creat} for both types of collagen fibers (oriented and nonoriented). The intensities of the DQF signal (at T=298 K) as a function of the creation time for C8y collagen fibers (oriented along the magnetic field B_0 and randomly oriented) are shown in Fig. (**4**).

Table 1. The fitting results (ω_L, Hz) of ^{1}H DQF NMR spectra for C8y and C15m collagen fibers oriented along the magnetic field B_0.

	C8y					C15m				
N	1	2	3	Mean value	σ	1	2	3	Mean value	σ
ω_L*	2200	2034	1748	1994	202	2018	1566	1436	1673	306
ω_L**	611	658	788	686	92	801	581	460	614	173
r^2	0.992	0.992	0.991			0.993	0.996	0.991		

The data are from the spectra displaced in antiphase, *i.e.*, negative (*) and positive (**) *Lorentz* peaks were estimated. The standard deviation (σ) and correlation coefficient (r^2) in each experiment at modeling line shapes of DQF spectra by two *Lorentz* functions [12 - 15] are displaced in the last column and last row, respectively.

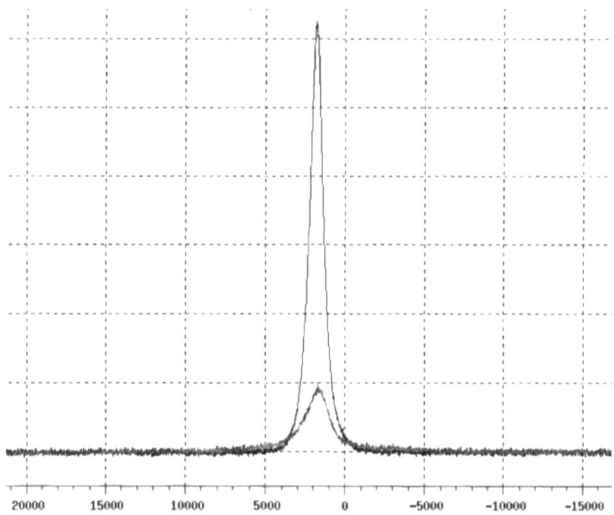

Fig. (3). ^{1}H DQF NMR spectra (intensity is in arbitrary units) of oriented fibers of C15m collagen (high intensity) and nonoriented fibers of C15m collagen (low intensity). T=298 K, and resonance frequency for protons is 400 MHz. Evolution time t_1=100 μs. The spectra have been recorded at $(\tau_{creat})_{max}$. Frequency scale is in Hz.

The dependencies of DQF signals (at T=298 K) on the creation time for oriented fibers of C8y and C15m collagens are compared in Fig. (**5**).

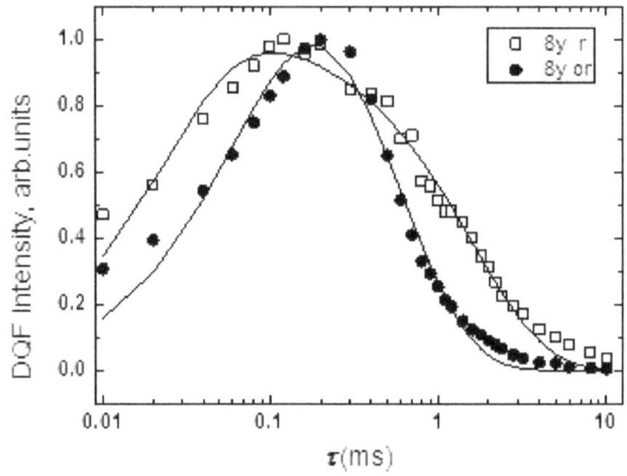

Fig. (4). ^{1}H DQF NMR signal intensity (normalized per maximal intensity) as a function of the creation time for randomly oriented collagen fibers C8y (□) and for collagen fibers C8y oriented along the magnetic field B_0 (●). T=298 K, frequency = 400 MHz, and HL=0.6. Solid lines are fits to the equation for transverse magnetization: $M_{xy}=A\times[\exp(-s_2\tau_{creat})-\exp(-s_1\tau_{creat})]$. Relaxation constants: s_1 =12.50 ms^{-1}, s_2 =1.78 ms^{-1} for oriented collagen fibers C8y (●) and s_1 =41.1 ms^{-1}, s_2 =0.617ms^{-1} for nonoriented collagen fibers C8y (□).

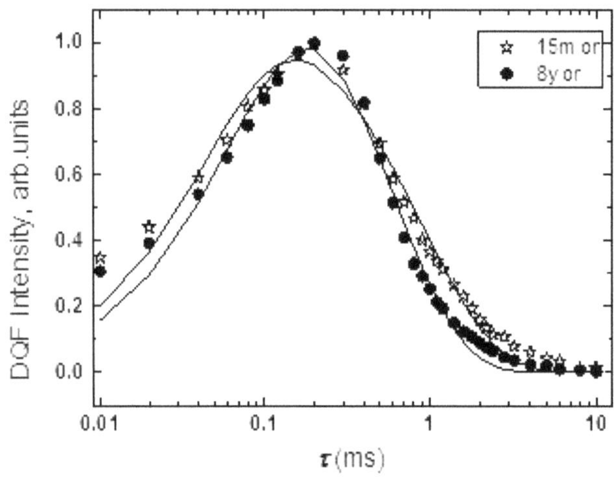

Fig. (5). ^{1}H DQF NMR signal intensity (normalized per maximal intensity) as a function of the creation time for C15m (*stars*) and C8y (●) collagen fibers oriented along the magnetic field B_0. T=298 K, frequency = 400 MHz, and HL=0.6. Solid lines are fits to the equation for transverse magnetization: $M_{xy}=A\times[\exp(-s_2\tau_{creat})-\exp(-s_1\tau_{creat})]$. Relaxation constants: s_1 =12.50 ms^{-1}, s_2 =1.78 ms^{-1} for collagen fibers C8y (●) and s_1 =19.92 ms^{-1}, s_2 =1.08 ms^{-1} for collagen fibers 15m (*stars*).

The DQF signals in oriented fibers reached the maximum intensities at creation times of $(\tau_{creat})_{max}$=0.165 ms and 0.18 ms for C15m and C8y collagen fibers, respectively. For randomly distributed collagen fibers, $(\tau_{creat})_{max}$ =0.100±0.005 ms (C8y) and $(\tau_{creat})_{max}$=0.101±0.016 ms (C15m) [1, 2]. According to the known results [3, 4, 6, 9, 10, 16], ^{1}H DQF NMR signals as a function of τ_{creat} are due to RDI (*i.e.*, to the degree of order in the sample) and the proton-exchange rate. The velocity of the exchange diminishes with decrease of temperature [3, 5]. The differences in the dependences of the DQF signals on the creation times could be observed for C8y and C15m collagens at decrease of temperature from 298 to 233 K [1]. Some examples of these comparative studies are presented in Fig. (**6**).

The data showed the increase of DQF signal when temperature decreased from 298 to 233 K. This increase is in accordance with the published data on DQF intensity obtained at changing temperature and thus depending on proton-exchange rate [4 - 6]. Fitting spectral lines (at 298 K and 233 K) by a difference of two Lorentzian functions also showed that the line widths of both functions increased with decreasing temperature. For instance, at 233 K for C8y collagen, the line width became ~7290 Hz (negative peak) and ~1730 Hz (positive peak), whereas for C15m fibers, it became ~3923 Hz (negative peak) and ~1214 Hz (positive peak), respectively.

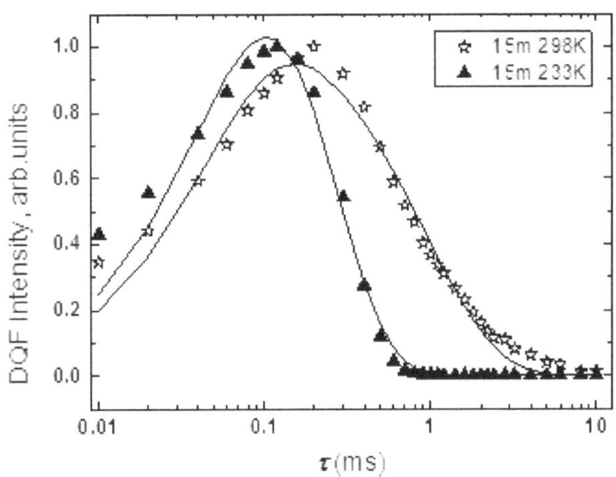

Fig. (6). ^{1}H DQF NMR signal intensity (normalized per maximal intensity) as a function of the creation time for C15m collagen fibers oriented along the magnetic field B_0 at T=298 K (*stars*) and T=233 K (▲). Frequency = 400 MHz, and HL=0.6. Solid lines are fits to the equation $M_{xy} = A\times[\exp(-s_2\tau_{creat})-\exp(-s_1\tau_{creat})]$. Relaxation constants: s_1 =19.92 ms^{-1}, s_2 =1.08 ms^{-1} for collagen fibers at T=298 K (*stars*) and s_1 =14.17 ms^{-1}, s_2 =6.18 ms^{-1} for collagen fibers at T=233 K (▲).

According to the DQF studies of tendon [9, 10], the shape of line and the amplitude of the ^{1}H DQF spectra are affected by the ratio of the residual ^{1}H–^{1}H dipolar interactions and the proton-exchange rate. Thus, the intensity of DQF signal declines when the proton-exchange rate increases. At studying tendon [9, 10], it was showed that the RDI is temperature independent. The proton-exchange rate changed by more than a factor of 3 at changing the temperature by 40° (from 313 to 274.5 K) [6, 9, 10]. In the DQF studies of C8y and C15m fibers [1, 2], the changes in temperature are larger (by 65°). Therefore, it is obvious that the effect of diminishing the proton-exchange rate has taken place in the increase of the DQF signal intensity at low temperature [1, 2].

At T=233 K, the DQF signals reached the maximum intensities at creation times $(\tau_{creat})_{max}$=0.10 and 0.07 ms for C15m and C8y collagens, respectively. The intensity of the peak in ^{1}H DQF NMR spectrum changed with decrease of temperature. These DQF studies discovered that, for instance, the intensity of the DQF peak in C15m collagen at $(\tau_{creat})_{max}$ was 1.27 times larger at 233 K than that at 298 K (at $(\tau_{creat})_{max}$), whereas in C8y collagen, the intensity of the peak at 233 K was 1.65 times larger than that at 298 K (at $(\tau_{creat})_{max}$). ^{1}H-^{1}H dipolar interaction depends on the order of water molecules interacting with the collagen fibers and the orientation of the fibers relative to magnetic field [13, 8 - 10, 16, 17].

Therefore, DQF studies have been also carried out on randomly oriented fibers of C15m and C8y for comparison with the data on fibers oriented along the direction of magnetic field B_0.

In oriented samples of C15m and C8y, the collagen fibers were located along the longitudinal axis of NMR tube. However, the fibers in nonoriented samples had mesh-like structures [2]. The experiments with those nonoriented fibers observed smaller contribution into the intensity of DQF signal (Fig. **3**). For example, for C8y collagen, a comparison of ^{1}H DQF NMR spectra showed that the line width of positive peak in random fibers (1640–1820 Hz) was larger than that in oriented fibers (900–1000 Hz). The intensity of DQF signal at $(\tau_{creat})_{max}$ for oriented fibers was 4.2 times larger than that for nonoriented one's in the case of C15m collagen. It was 4.8 times larger in the case of C8y fibers [1]. This relation changed in the range of 3.0–7.0 for different samples with the same value of HL [1, 2].

The data obtained in DQF studies demonstrated that it is possible to detect ^{1}H DQF NMR signal of water in C8y and C15m collagen fibers with hydration level of 0.6 g H_2O per g dry mass that results from RDI [1]. C8y and C15m collagen fibers have different values of $(\tau_{creat})_{max}$. The difference in $(\tau_{creat})_{max}$ between two collagens with different cross-links is not so big (*e.g.*, 0.16 ms for C15m and 0.186 ms for C8y in one comparative experiment). However, it is in agreement with the published data [3, 5, 9] that different collagen containing tissues are characterized by the different dependence of DQF signal on the DQ creation time and the different value $(\tau_{creat})_{max}$.

The ^{1}H DQF NMR technique filters out contributions from free water molecules. As a result, it enables the comparison of adult (C8y, higher level of cross-links) and young (C15m, low cross-link level) collagen fibers according to the measurements of NMR properties of water molecules in the vicinity of anisotropic sites. The detection of ^{1}H DQF NMR signal indicates the presence of anisotropy (order) in collagen tissues. Moreover, this order is higher in collagen with highest cross-link level because the intensity of ^{1}H DQF signal in C8y is 1.8 times larger than the one in C15m collagen. With age, cross-links in collagen progress from immature to more mature forms. The studies also indicated that the orientation of collagen fibers is the main factor in the determination of the order of water molecules. Therefore, this results in the increased intensity of DQF signal [1, 3]. The order of water molecules induced by the collagen fibers depends on the degree of the order of fibers. If the collagen fibers are in the sample in mesh-like arrangement (as in the case of randomly oriented fibers), then dipolar interactions are largely averaged resulting in weaker RDI. Hence, weaker DQF signal intensity will be observed in comparison with that in oriented collagen fibers. The samples with nonoriented fibers (both in C8y and C15m) contain numerous collagen fibrils, but they are not well aligned with the longitudinal axis of NMR tube. The results on C15m and C8y collagens are in accordance with the published data for such collagen containing tissues, which contain different level of collagen molecules, and which are aligned well along the fiber director or have mesh-like arrangement. For example, a comparison could be done between skin (the collagen content of 20%) and tendon (the collagen content of 30%) [3, 5, 9]. It was shown that, if the collagen fibers run randomly or have mesh-like structures (*e.g*, skin), the intensity of ^{1}H DQF signal is weaker relative to the DQF signal in the tendon (fibers are well oriented along the longitudinal axis). Similar to the collagen fibers in tendon, there is a macroscopic order of water in oriented C8y and C15m fibers. The presence of this macroscopic order in studied adult and young collagen fibers at HL = 0.6 is in accordance with the published data [9, 17, 18] on anisotropic motion of water molecules in hydrated (HL = 0.45) collagen, which have been characterised *via* the average orientation of water molecules with respect to the fiber axis [16, 17]. The experiments with oriented and nonoriented C8y and C15m fibers showed that, when we change the oriented fibers (with macroscopic order of water in them) with the nonoriented ones, a macroscopic order is indeed destroyed, and the intensity of ^{1}H DQF NMR signal is decreased, basically, circa in 4 times. The result values scattered from 3 to 8 (for physically separate samples) [1]. Thus, an observation of the ^{1}H DQF NMR signal of smaller intensity in random C8y and C15m fibers testified that local order (associated with water within the microfibrils of collagen fibrils and the interfibrillar spaces [19, 20]) results in DQF signal also. This is in accordance with the description [5] that the residual dipolar interaction enabling to get the ^{1}H DQF NMR signal may be of local or macroscopic nature. Thus, the ^{1}H DQF NMR data on randomly oriented fibers of C8y and C15m collagens (in comparison with oriented ones) give a possibility to detect local order and to take that into account at analysis of macroscopic order fraction from DQF signal of oriented collagen fibers.

The larger value of line width in nonoriented C8y and C15m collagen fibers (approximately two times) also indicates the dependence of this characteristic on the fiber orientation. In the presence of macroscopic order, the width of the DQF spectra should follow the factor ($3\cos^2\theta-1$), where θ is the angle between the magnetic field and the director of the average RDI over the whole sample [3, 5]. In the case of alignment of the direction of the dipolar interaction along the collagen fibers, a change in the fiber orientation from parallel to the perpendicular position relative to the external magnetic field will result in a change of line width. However, the collagen samples studied were heterogeneous [1], and the line width varied for different creation times. DQF studies on tendons [5, 9, 10] also paid attention to this effect, that is, the width of the line was depending on the factor ($3\cos^2\theta-1$).

Thus, ^{1}H DQF NMR spectroscopy is able to observe the RDI in water molecules in oriented and randomly oriented collagen fibers from connective tissues of different age animals (15-month-old calf and 8-year-old steer) at water content of 0.6 g H_2O per g dry matter. The quantity of covalent intermolecular cross-links in collagens increases during the life of animal. It was observed that intensities of DQF signals in fibers oriented along outer static magnetic field exceeded significantly the DQF signals in randomly oriented fibers, *i.e.*, without determined orientation with respect to magnetic field (Fig. **3**). The results can be analysed on the basis of tensor description of a DQF signal formation. Special sequence of RF pulses with suitable phase cycling was used (Chapter 1, Fig. **9**) as in the studies of other systems with anisotropic water mobility [5 - 7]. The dependency of DQF signal upon creation time (τ_{creat}) forming the ($T_{2,+1}$, $T_{2,-1}$) second rank from ($T_{1,+1}$, $T_{1,-1}$) tensors was described by the difference of two exponential functions. The values of $(\tau_{creat})_{max}$ were compared for both collagens with different quantities of intermolecular cross-links. The dependence of the DQF signal on the creation time of the DQ coherences is often characteristic to each tissue.

The RDI may be of local or macroscopic nature. In both the cases, DQF spectra are detected. In the latter case, the spectra would change when the sample is rotated. In the cases of anisotropic molecular tumbling, the averaging of the dipolar interaction is partial. Due to the fact that the ^{1}H DQF NMR technique filters out contributions from free water molecules, it enables us to specifically measure NMR properties of water molecules in the vicinity of anisotropic sites.

^{1}H DQF NMR STUDY OF *BOMBYX MORI* SILK

An interaction of some natural fibrous materials with water leads to a decrease in the length of the fibers [13, 21]. The role of water is not completely clear in structure and properties of some fibrous biopolymers, in particular, various types of natural silks [12, 22, 23]. For example, natural silk produced by the silkworm *B. mori* is known as one of the strong natural and synthetic materials [13, 22 - 24]. Natural silk *B. mori* with low water content has been studied by ^{1}H DQF NMR and single-pulse ^{1}H NMR methods [2, 7, 12, 13, 23]. These studies demonstrated that DQF NMR enables one to probe the anisotropic motion of water in the silk fibers with low water content. DQF spectroscopy uses a special sequence of RF pulses with phases, which are cyclically varied to suppress undesirable signals, in particular, the signals of isotropic water [3, 6, 13]. In an isotropic medium, the dipole interaction is completely averaged to zero. Therefore, DQF signal emerges only if there is ordered anisotropic structure in the sample. In particular, ^{1}H DQF NMR spectra were recorded for many tissues (tendon, ligament, skin, *etc.*) which contain collagen [1, 3, 4, 7]. DQF signals of anisotropic water are generated in fibrous systems as a result of the interaction with collagen fibers. One can assume that water bound to silk fibers also exhibits anisotropic properties, for example, RDI. Various processes induced by the pulse sequence presented in Chapter 1, Fig. (**9**), are described in terms of the representations of spherical tensor operators $T_{l,p}$, where l is the tensor rank and p (coherence) can be varied by a nonselective radio frequency pulse within the range of $-l \leq p \leq +l$ [1, 12, 13]. In an anisotropic medium, where dipole interactions are not averaged to zero, even rank tensor (*e.g.*, $T_{2,+1}$ and $T_{2,-1}$) can be created within creation time τ_{creat}. The detected DQF signal of water in *B. mori* natural silk is a result of the anisotropic molecular motion. The characteristic ^{1}H DQF NMR spectrum has been recorded for the sample with HL=0.065 [13]. For moist sample with HL=0.18, no DQF signals were observed. For silk sample with HL=0.065, the shape of the line of the DQF NMR signal was described by a combination of a narrow *Lorentzian* line (ω_{L1} =5136 Hz) and a broad *Lorentzian* line (ω_{L2} =20070 Hz). The phase of the first line was opposite to that of the second. The attempts to model this DQF spectrum by a single *Lorentzian* line (ω_L =4850 Hz) gave worse results. This suggests that for the natural silk fibers, the studied two lines (narrow and broad) were combined in antiphase. An estimation of the line width could allow one to compare the two relaxation constants s_1 and s_2 in expression for transverse magnetization described in DQF experiment in Chapter 1, Fig. (**9**) [13, 15]. The results on randomly distributed silk fibers (HL=0.065) showed that the DQF signal intensity *versus* creation time decreased with increasing creation time (Fig. **7**) [13]. This signal is induced by residual dipolar interactions undetectable at higher water content (HL=0.18). Since ^{1}H DQF NMR method filters the contributions of isotropic water, one can consider that the properties of silk with low water content are due

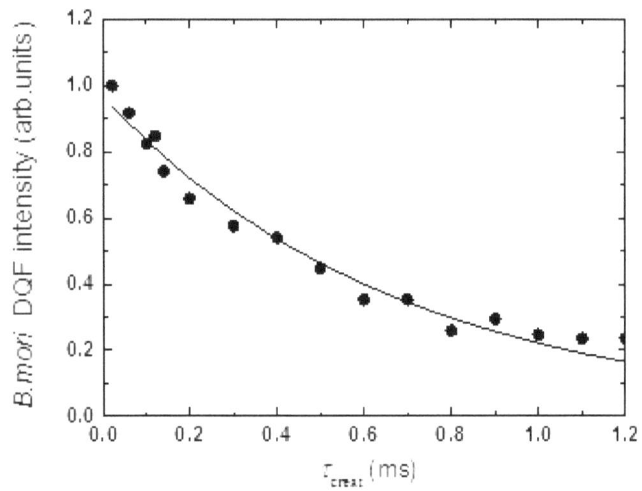

Fig. (7). ^{1}H DQF NMR signal intensity (normalized per maximal intensity) as a function of the creation time for randomly oriented *B. mori* silk fibers. T=298 K. The proton frequency is 400 MHz. HL=0.065. Solid line is the fit to exponential function.

to the properties of water molecules near anisotropic sites. ^{1}H DQF NMR signal is an indicator of order in the natural silk samples studied. Two cases are normally considered at registration of the second rank tensors ($T_{2,+1}$ and $T_{2,-1}$) by MQF NMR spectroscopy. These are anisotropic motions in the systems with either macroscopic or local order. In the first case, the directors have a single spatial orientation (as it was in oriented collagen fibers or tendons) [1, 3, 5, 9], whereas in the second case, the orientations of local directors are randomly distributed. In the studied samples of randomly distributed silk fibers, there was no single spatial orientation, that is, no macroscopic order [12, 13, 23]. The obtained ^{1}H DQF NMR signal of *B.mori* fibers was weak (large acquisitions had to be used). In the absence of selected orientation of fibers in the silk samples with HL=0.065, the observation of DQF signal is a confirmation of the local order [13]. In some studies, such an anisotropic interaction resulted in RDI, which is nonaverageable to zero, and in splitting of resonance lines [3, 10, 16, 17]. The ability to detect such splitting depends on the proportion between the isotropic and anisotropic signals, the degree of splitting, and other factors [1, 3]. In some biological tissues, the splitting is small because of a low fraction of bound water and a small value of the order parameter of water molecules bound to oriented biopolymers. In this case, the splitting is mostly hidden by the signal of isotropic water. Although the problem of RDI registration is partially solved by MQF methods [3 - 9], a proton doublet (produced by splitting) at temperature above 298 K merges into singlet [5, 9]. This is the sign of proton exchange between water molecules in anisotropic and isotropic states. It has been shown that if RDIs (ω_{H-H}) are weak (*i.e.*, rate constant $k >> 3\omega_{H-H}$), the intensity of the DQF spectrum is proportional to $1/k^2$ [3, 5]. This is likely to be a reason why at room temperature (when k is already quite substantial) [3, 10], only a single weak resonance line is detected in NMR spectrum of natural silk. The absence of a DQF signal in the silk with HL=0.18 [13] can be caused both by large k value and by the ability of water to plasticize silk. In other words, additional humidification of silks can even make some amino acids as isotropically mobile (on NMR scale) [12, 13]. In single-quantum NMR experiments with *B.mori* silk, the line widths (the parameters of *Lorentzian* line ω_L with *Gaussian* broadening ω_G) vary with change in water content. Thus, it indicates an increase in the mobility of water molecules with an increase in water content from 0.065 to 0.18. It has been emphasized [3] that the fraction of mobile isotropic water always interfered with the signal of anisotropic water molecules [16, 17, 19].

According to some data [3, 5, 10], the DQF signal curves *versus* τ_{creat} are regarded as the characteristics of various tissues. Moreover, the values of creation times at which DQF signal is maximal are also used to characterise different tissues. The data on some tissues (*e.g.,* brain tissue) showed that the DQF signal intensity only decreased with increasing τ_{creat} [3, 10]. For *B.mori* silk, only a decrease in the DQF signal intensity with an increase in τ_{creat} has been observed [13]. The result for the brain tissue was explained by the very fast creation of the DQF signal during the pulses. Therefore, this tissue was characterised by large line width and short relaxation time. In comparison with these data, the results on natural silk fibers discovered much broader lines and fast creation of the DQF signal at short τ_{creat}. This led to the absence of a maximum in the dependency $I_{DQF} = f(\tau_{creat})$ and caused a continuous decrease in I_{DQF} with increasing τ_{creat} [13] (Fig. **7**).

I_{DQF} in *B.mori* natural silk decreased and attained a value of 20% (or less) of the maximal intensity in experiments with $\tau_{creat} \geq 0.01–0.02$ ms at $\tau_{creat} \approx 0.8–1.0$ ms. The DQF method detects only anisotropic signals. In *B.mori* natural silk, the anisotropic motion is observed only at low water content (*e.g.,* HL=0.065). At this humidity level, the mobility of water molecules is characterized by correlation time 2.3×10^{-7} s. This motion is absent at HL≥0.18, when the correlation times are $\approx 1 \times 10^{-7}$ s or shorter.

Thus, the ^{1}H DQF NMR studies of *B.mori* natural silk fibers with low water content detected slow motions of water molecules and anisotropic behavior. The detected ^{1}H DQF NMR signals characterised the local order and RDI. It was found that the intensity of the DQF spectrum depends on the creation time. This is connected with forming second-rank tensors $T_{2,+1}$ and $T_{2,-1}$ [1, 23]. A possibility to observe the ordered water fraction in natural biological tissues resulted in an additional approach to characterise the molecular mobility of water in *B.mori* natural silk fibers with low water content.

CONCLUDING REMARKS

The technique of ^{1}H double-quantum-filtered (DQF) NMR spectroscopy has been considered in studying anisotropic motion of water in hydrated fibers of collagens and natural silk. The results were analyzed on the base of tensor description of DQF signal formation at the action of special sequence of radio frequency pulses. ^{1}H DQF NMR signals were responsible for RDI between protons of water molecules interacting with oriented fibers of natural polymers. It was considered that the dependences of DQF intensity *versus* τ_{creat} are the characteristics of various tissues and materials informing molecular order on macroscopic and microscopic scales [4 - 6, 25]. The technique is sensitive enough to depict many differences in natural materials including changing cross-links with aging.

REFERENCES

[1] V.V. Rodin, L. Foucat, and J.P. Renou, "The dipolar interactions and dynamics of water molecules in collagen fibers from two connective tissues of different ages by ^{1}H double-quantum-filtered NMR spectroscopy", *Mater. Sci.,* no. 5, pp. 2-10, 2004.

[2] V.V. Rodin, "Methods of Magnetic Resonance in Studying Natural Biomaterials", In: *Encyclopedia of Physical Organic Chemistry,* Zerong Wang, Ed., Chapter 53. vol. 4, part 4. John Wiley & Sons, Inc., 2017, pp. 2861-2908.

[3] G. Navon, H. Shinar, U. Eliav, and Y. Seo, "Multiquantum filters and order in tissues", *NMR Biomed.,* vol. 14, no. 2, pp. 112-132, 2001. [http://dx.doi.org/10.1002/nbm.687] [PMID: 11320537]

[4] U. Eliav, M. Komlosh, P.J. Basser, and G. Navon, "Characterization and mapping of dipolar interactions within macromolecules in tissues using a combination of DQF, MT and UTE MRI", *NMR Biomed.,* vol. 25, no. 10, pp. 1152-1159, 2012. [http://dx.doi.org/10.1002/nbm.2784] [PMID: 22362643]

[5] U. Eliav, and G. Navon, "A study of dipolar interactions and dynamic processes of water molecules in tendon by ^{1}H and ^{2}H homonuclear and heteronuclear multiple-quantum-filtered NMR spectroscopy", *J. Magn. Reson.,* vol. 137, no. 2, pp. 295-310, 1999. [http://dx.doi.org/10.1006/jmre.1998.1681] [PMID: 10089163]

[6] G. Navon, U. Eliav, D.E. Demco, and B. Blümich, "Study of order and dynamic processes in tendon by NMR and MRI", *J. Magn. Reson. Imaging,* vol. 25, no. 2, pp. 362-380, 2007. [http://dx.doi.org/10.1002/jmri.20856] [PMID: 17260401]

[7] V.V. Rodin, "Translational dynamics of molecules in synthetic and natural polymers as studied by NMR", *27th ECIS Conference,* 2013 September 1-6, Sofia, Bulgaria.

[8] V.V. Rodin, "NMR approaches in studying dynamic molecules in porous materials", *J. Mater. Sci. Eng.,* vol. 3, no. 3, p. 245, 2014.

[9] Y. Seo, K. Ikoma, H. Takamiya, Y. Kusaka, L. Tsoref, U. Eliav, H. Shinar, and G. Navon, "^{1}H double-quantum-filtered MR imaging as a new tool for assessment of healing of the ruptured Achilles tendon", *Magn. Reson. Med.,* vol. 42, no. 5, pp. 884-889, 1999. [http://dx.doi.org/10.1002/(SICI)1522-2594(199911)42:5<884::AID-MRM7>3.0.CO;2-0] [PMID: 10542346]

[10] L. Tsoref, H. Shinar, Y. Seo, U. Eliav, and G. Navon, "Proton double-quantum filtered MRI-A new

method for imaging ordered tissues", *Magn. Reson. Med.,* vol. 40, no. 5, pp. 720-726, 1998. [http://dx.doi.org/10.1002/mrm.1910400512] [PMID: 9797155]

[11] C.R. Vanderschee, and K.J. Ooms, "Investigating water interactions with collagen using ^{2}H multiple quantum filtered NMR spectroscopy to provide insights into the source of double quantum filtered signal in tissue", *J. Phys. Chem. B,* vol. 118, no. 13, pp. 3491-3497, 2014. [http://dx.doi.org/10.1021/jp409543p] [PMID: 24628019]

[12] V.V. Rodin, G.M. Reznichenko, and E.L. Vasina, "Properties of natural polymer fibers with low water content", *Mater. Sci.,* no. 1, pp. 34-42, 2004.

[13] V.V. Rodin, and D.P. Knight, "Molecular mobility in natural polymers: *Bombyx mori* silk with low water content as studied by ^{1}H DQF NMR", *Biophysics,* vol. 49, no. 5, pp. 730-737, 2004. [PMID: 15526463]

[14] R.C. Lyon, J. Pekar, C.T. Moonen, and A.C. McLaughlin, "Double-quantum surface-coil NMR studies of sodium and potassium in the rat brain", *Magn. Reson. Med.,* vol. 18, no. 1, pp. 80-92, 1991. [http://dx.doi.org/10.1002/mrm.1910180110] [PMID: 2062244]

[15] J. Pekar, and J.S. Leigh, "Detection of biexponential relaxation in sodium-23 facilitated by double-quantum filtering", *J. Magn. Reson.,* vol. 69, no. 3, pp. 582-584, 1986.

[16] C. Migchelsen, and H.J.C. Berendsen, "Proton exchange and molecular orientation of water in hydrated collagen fibers. An NMR study of H_2O and D_2O", *J. Chem. Phys.,* vol. 59, no. 1, pp. 296-305, 1973. [http://dx.doi.org/10.1063/1.1679805]

[17] J.R. Grigera, and H.J.C. Berendsen, "The molecular details of collagen hydration", *Biopolymers,* vol. 18, pp. 47-57, 1979. [http://dx.doi.org/10.1002/bip.1979.360180106]

[18] H.J.C. Berendsen, Specific interactions of water with biopolymers.*Water, A Comprehensive Treatise.,* F. Frank, Ed., vol. 5. Plenum Press: New York, 1975, pp. 293-330.

[19] A. Traore, L. Foucat, and J.P. Renou, "^{1}H-NMR study of water dynamics in hydrated collagen: transverse relaxation-time and diffusion analysis", *Biopolymers,* vol. 53, no. 6, pp. 476-483, 2000. [http://dx.doi.org/10.1002/(SICI)1097-0282(200005)53:6<476::AID-BIP4>3.0.CO;2-8] [PMID: 10775063]

[20] J.Y. Exposito, U. Valcourt, C. Cluzel, and C. Lethias, "The fibrillar collagen family", *Int. J. Mol. Sci.,* vol. 11, no. 2, pp. 407-426, 2010. [http://dx.doi.org/10.3390/ijms11020407] [PMID: 20386646]

[21] J.D. van Beek, J. Kümmerlen, F. Vollrath, and B.H. Meier, "Supercontracted spider dragline silk: a solid-state NMR study of the local structure", *Int. J. Biol. Macromol.,* vol. 24, no. 2-3, pp. 173-178, 1999. [http://dx.doi.org/10.1016/S0141-8130(98)00083-X] [PMID: 10342762]

[22] G.J. Davies, D.P. Knight, and F. Vollrath, "Chitin in the silk gland ducts of the spider Nephila edulis and the silkworm *Bombyx mori*", *PLoS One,* vol. 8, no. 8, p. e73225, 2013. [http://dx.doi.org/10.1371/journal.pone.0073225] [PMID: 24015298]

[23] V.V. Rodin, G.M. Reznichenko, and E.L. Vasina, "Properties of natural silk fibers: deformation studies and NMR data", *Biophysics,* vol. 49, no. 6, pp. 918-926, 2004. [PMID: 15612542]

[24] Os. Hakimi, D. Knight, F. Vollrath, and P. Vadgama, "Spider and mulberry silkworm silks as compatible biomaterials", *Compos., Part B Eng.,* vol. 38, pp. 324-327, 2007. [http://dx.doi.org/10.1016/j.compositesb.2006.06.012]

[25] G.S. Payne, and P. Styles, "Multiple-quantum-filtered ^{23}Na NMR spectroscopy in model systems", *J. Magn. Reson.,* vol. 95, no. 2, pp. 253-256, 1991.

CHAPTER 5

NMR Relaxation and Restricted Self-Diffusion of Water in Wood

Abstract: In the technological treatment of wood materials, it is important to control water-wood interactions. To measure the water self-diffusion coefficients properly in wetting/drying processes of wood, it is necessary to understand many factors affecting the experiments. For fibrous material, the diffusivities of water could be different along the fiber and in the direction of perpendicular to fiber axis. The chapter presents the data of NMR relaxation and 1-dimensional PFG NMR in studying anisotropic diffusion of water in wood.

Keywords: Apparent diffusion coefficient D_{app}, Cross-relaxation (CR), Diffusion anisotropy, Free induction decay (FID), Goldman-Shen (GS) experiment, Inverse Laplace transform (ILT), Inversion recovery (IR) method, NMR, Pulsed field gradient (PFG) NMR, Spin-lattice (longitudinal) relaxation time T_1, Spin-spin (transverse) relaxation time T_2, Stimulated echo (STE), Water, Wood.

INTRODUCTION: WATER IN WOOD

Wood is a well known natural fiber composite [1 - 3]. Moisture and temperature affect the properties of wood [2, 4 - 6]. Wetting and drying wood are the most natural factors defining the properties of wood suitable for further manufacturer treatment [1, 2, 4]. In order to regulate a diffusion of water in wood materials properly, it is necessary to know water dynamics at wetting/drying wood [4 - 6]. Model and experimental studies should be developed to allow operators and manufacturers of wood structures to utilize nondestructive techniques to evaluate and accurately identify technological parameters for structural performance [2, 7].

^{1}H NMR is a unique nondestructive technique which provides the information about the interactions between wood and water in materials [3 - 6]. The NMR methods lead to determine the mobility of the water in the porous network of materials by relaxation times/diffusion coefficients [6, 8 - 16]. The experimental data (magnetization decays) are often fitted by a sum of several relaxation components or are presented by a distribution of NMR relaxation times.

Victor V. Rodin

The following sections consider wood-water interactions analyzing the data obtained from ^{1}H NMR relaxation and ^{1}H NMR PFG diffusometry in one-dimension.

STUDYING NMR RELAXATION IN WOOD: FROM WET TO DRIED WOOD

Wood samples were taken from the pieces of felled mature spruce that had been kept frozen prior to use. The wood samples for NMR measurements were prepared from fresh pieces of wood and kept covered in the polymer film in the fridge (278 K) before and between the experiments. Wood samples were cut from the original pieces of wood as follows: in the first case, the samples were roughly right cube size (6×6×6 mm^3) that allowed the rotation of the wood piece by 90° in NMR tube leaving the direction of gradient to main magnetic field as constant; in the second case, the samples were of parallelepiped size (6×6×10 mm^3). Each of the pieces was then placed in 10 mm NMR tube and sealed. Pulsed field gradient (PFG) NMR studies have been carried out on wood samples with humidity level (HL) higher than fiber saturation point (FSP). HL of wood was expressed as g H_2O per g dry wood mass [2, 4]. The FSP is the point at which all free water has been removed from the wood, and there was only bound water left. Normally, this happens at HL = 0.25-0.30 [4 - 8, 10] Above the FSP, in addition to the components of spin-spin relaxation times T_2 (between 0.1 and 10 ms), a slow relaxing T_2 component (in the range of 10-150 ms) then could be observed. At least three T_2 components in total could be observed in the wet wood above the FSP [4, 5, 8, 10].

Fig. (**1**) shows the result of the single-pulse ^{1}H NMR experiment (400 MHz, 298 K) when the FID signal after 90° pulse (length of the pulse = 8-10 μs) in hydrated wood sample has been measured. This FID was converted also from the time-domain to the frequency-domain using the FT. The result of this transformation presented a spectral line of proton resonance signal. The signal of total time-domain in hydrated wood sample (Fig. **1**) is the sum of the signals of the water protons and the protons of macromolecules of wood: $S = S_{wat} + S_{wood}$. The wood part of the FID was fitted by the Gaussian-sinc function [8, 14], *i.e.*, $S_{wood} = S_{wood(0)} \cdot \exp(-a^2t^2/2) \cdot \mathrm{sinc}(bt)$, where $S_{wood(0)}$ is the intensity of protons of wood at t = 0, and *a, b* are the parameters which are used in the equation of the second moment of the line width of M_2 [8, 14] as shown in eq. (1):

$$M_2 = a^2 + \frac{b^2}{3} \tag{1}$$

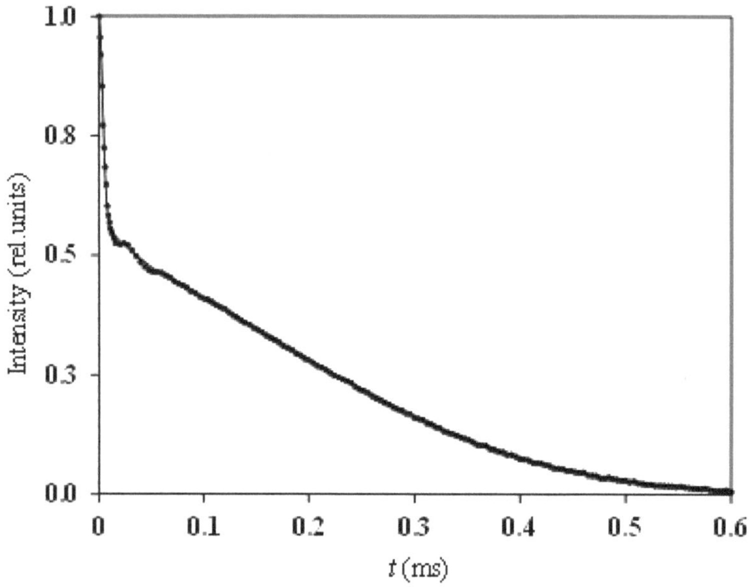

Fig. (1). Characteristic FID signal of protons in wood (HL=0.0303). T = 298 K, and a frequency is 400 MHz. Solid line is the fit to the experimental data as the sum of the signals of water protons and the protons of wood [8, 14]. The fast relaxing part of the FID (the signal of wood protons) was fitted by the Gaussian-sinc function [8], and slowly relaxing protons component from the total FID signal (water protons) was fitted by the Voigt function [14, 17].

The contribution of protons of macromolecules of wood should be avoided if there is a task to model the water signal only. Such approach was usually used when only the investigation of water in hydrated wood/other fiber materials was of main interest [8, 17, 19]. For this case, the measurements were carried out at experimental conditions when transverse magnetization of macromolecule protons decays very fast to zero (whereas the transverse water magnetization is merely partially decayed) and only protons of water could be measured in the study. The spin-lattice relaxation time T_1 was measured by the IR method using the pulse sequence (180°–τ–90°) at the pulse sequence repetition time $\geq 5T_1$ [8, 14, 18]. The spin-spin relaxation time T_2 in the wood was measured using the CPMG pulse sequence [8, 13, 14] at a time of 0.025 ms between π pulses. Normally, 8000 echoes have been applied to cover all range for CPMG decay in wet wood. For dry wood samples, about 1000 echoes were often enough to perform correct CPMG experiment (Chapter 1, Fig. **3**). The analysis of relaxation data was conducted after transferring raw data from spectrometer computer to desk computer using *MatLab* codes and 1D ILT software to present the data as T_2 distribution. One example of such distribution of the spin-spin relaxation times is shown in Fig. (**2**).

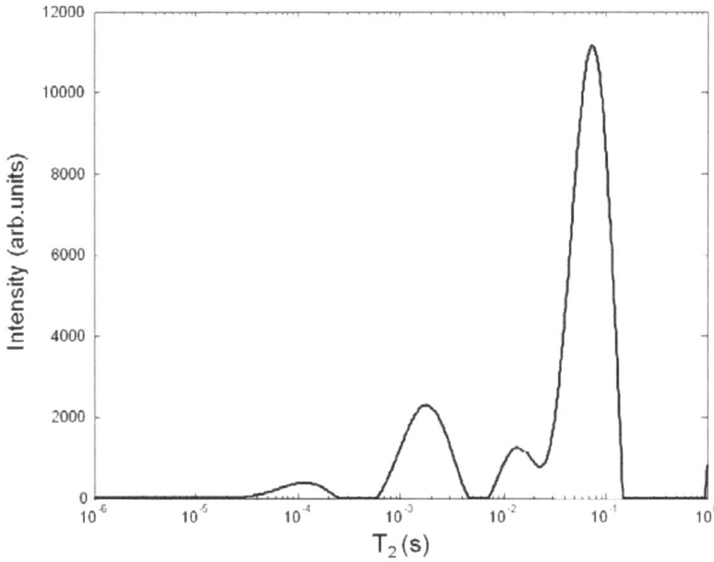

Fig. (2). T_2 distribution of ^{1}H obtained by 1D ILT treatment of CPMG data set measured on wet wood sample (HL=1.303) at 298 K, and a frequency of 400 MHz. *X*-axis is presented in log-scale from 10^{-6} s to 1 s. The integral intensities of the peaks (*from left to right*) are 3.1%, 17.4%, 79.5% of measurable protons. The last broad peak with the area of 79.5% in the range of T_2 = 8-150 ms is combined from two peaks of 6.9% and 72.6% population indeed.

There is a proton exchange between the water and exchangeable protons of macromolecular pool, and it is associated with an increase in the intensity of slow relaxing T_1 component [9, 14, 17, 18]. This CR effect was observed earlier in many macromolecular systems with low water content [8, 9, 17 - 21]. CR experiments on wood samples were implemented according to the *Goldman-Shen* (GS) pulse sequence ($90°_x – \tau_0 – 90°_{-x} – t – 90°_x$) which was applied before in the CR studies of other hydrated systems [8, 17, 18, 20, 21]. The first two pulses are used to create the different conditions for longitudinal magnetizations of water protons and collagen protons. The time gap τ_0 is varied to let the wood magnetization disappears but leave the water magnetization. The second pulse (with opposite phase to the first one) rotates water proton magnetization back to the steady magnetic field. In the case of effect of cross-relaxation, longitudinal macromolecule and water magnetization will change during the time *t*. After the third $90°_x$ pulse, a measurement of free induction decay of water protons has been done. Fig. (**3**) shows the typical GS experimental curves recorded for wood sample with HL=0.03 g H_2O per g dry mass varying τ_0.

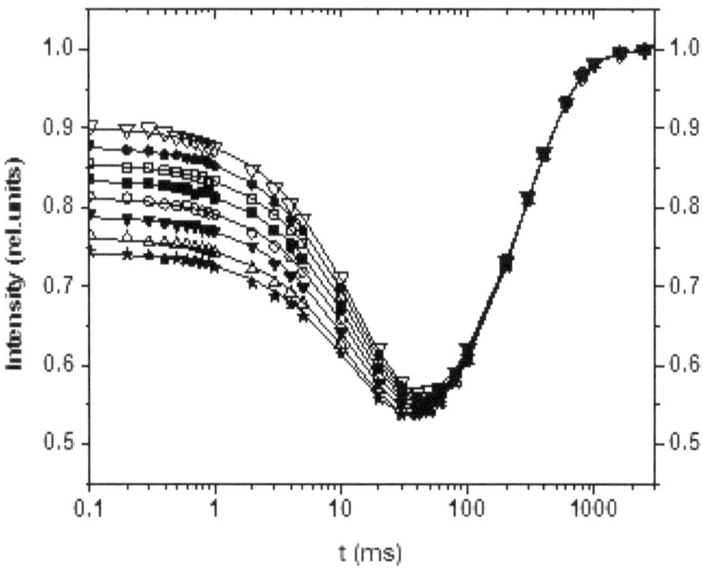

Fig. (3). The intensities of proton signals of water as a function of interval t (spacing between the 2nd and 3rd 90° pulses in Goldman-Shen pulse sequence $90°_x–\tau_0–90°_{-x}–t–90°_x$) for the wood samples at HL = 0.03 g H_2O per g dry mass varying τ_0 from 65 μs (*top curve*) to 205 μs (*bottom curve*) with step of 20 μs. Solid lines are the fits of Equations (2)-(7) to the experimental data according to the theory [8, 9, 20]. The dip on the curve shows the existence of a cross-relaxation. T= 298 K, and a frequency is 400 MHz.

PFG NMR studies have been realized in one-dimension case by the three pulse stimulated-echo (STE) sequence [6, 8, 9, 13, 14, 22] as described with details in the chapter 1, Fig. (7), *top*. The 90° radio frequency excitation pulse was established as 10 μs. In order to set appropriate repetition time in the diffusion experiment, the T_1 values of the water signal from each sample were estimated by IR method [9, 14]. Doped water sample (with $MnCl_2$) was prepared and measured in similar NMR experiments to test all pulse sequences and clarify experimental parameters for the studies of wood [8, 9].

According to the published works [8, 9, 14, 17], the amplitude of echo depends on the gradient factor $b = q^2(\Delta - \delta/3)$ and the echo time $(\tau_2 + 2\tau_1)$ as: $I(\tau_1,\tau_2,b)/I(0) = \exp(-\tau_2/T_1)\cdot\exp(-2\tau_1/T_2)\cdot\exp(-b\cdot D)$. So, for STE experiment, the additional echo-attenuation, $I(G)/I^*(0)$, is given in chapter 1, eq.(6). Originally, raw data set in diffusion experiment with gradient in one direction was recorded as a number of echo signals in the frequency domain at varying gradient values. Fig. (**4**) shows typical 3D plot taken in diffusion experiment on particular wood sample. Further, the data sets were treated according to the procedures and methods described earlier in the chapter 1.

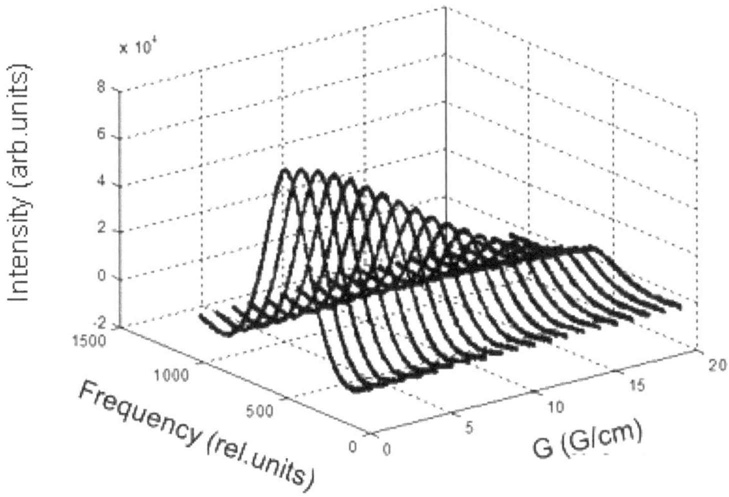

Fig. (4). Typical 3D plots for raw data on wood sample measured on CMX-400. Echo-attenuation is presented for diffusion time Δ=70 ms. T = 298 K, and a frequency is 400 MHz. HL=0.9.

RESTRICTED DIFFUSION OF WATER IN WOOD: ONE-DIMENSIONAL PFG NMR STUDIES

In STE diffusion experiment, longitudinal relaxation has a place during diffusion interval between the 2nd and 3rd RF pulses (Chapter 1, Fig. **7**, *top*). The rate of cross-relaxation characterizing magnetization transfer between both phases (water and wood) is a function of q. It will consequently depend on gradient G. A behaviour of Z-magnetization of water protons and that of macromolecules can be analysed with the aid of *Bloch* equations with exchange terms [17, 18, 20]. Ref. [20] shows how the extended *Bloch* equations describe magnetization of both phases, and how the STE signal is affected by cross-relaxation and diffusion. CR influences only on longitudinal magnetization [18 - 21]. Therefore, Ref. [20] presented a description of magnetizations of water and macromolecule protons during τ_2, *i.e*, in conditions of longitudinal relaxation (Chapter **1**, Fig. **7**, *top*) [9, 17, 20].

The data were analysed in terms of the two-phase model [8, 13, 17 - 20]. R_{1w} and R_{1m} are the intrinsic longitudinal relaxation rates, k_w and k_m are the cross-relaxation rates, p_w and p_m are the fractions of the protons belonging to the phases of water and macromolecular system, respectively. $p_w+p_m=1$. We consider $M_w(t)$ as the longitudinal magnetization of water pool. The time dependent normalized deviation $\Delta m_w = (M_w(t)-M_w^{eq})/M_w^{eq}$ from the equilibrium value of longitudinal magnetization of the water pool (M_w^{eq}) can be presented then according to [9, 18, 20], *i.e.* in the following eq.(2):

$$\Delta m_w = \frac{M_w(t) - M_w^{eq}}{M_w^{eq}} = c^+ e^{-R^+ t} + c^- e^{-R^- t} \tag{2}$$

The parameters of c^+, c^-, R^+, R^- are connected to spin-lattice relaxation rates of R_{1w} and R_{1m} and cross-relaxation rates of k_w and k_m according to the equations (3)-(6):

$$2R^+ = k_w + R_{1w} + k_m + R_{1m} + [(k_w + R_{1w} - k_m - R_{1m})^2 + 4k_w k_m]^{1/2} \tag{3}$$

$$2R^- = k_w + R_{1w} + k_m + R_{1m} - [(k_w + R_{1w} - k_m - R_{1m})^2 + 4k_w k_m]^{1/2} \tag{4}$$

$$c^+ = +\Delta m_w(\tau_0)\frac{k_w + R_{1w} - R^-}{R^+ - R^-} - \Delta m_m(\tau_0)\frac{k_w}{R^+ - R^-} \tag{5}$$

$$c^- = -\Delta m_w(\tau_0)\frac{k_w + R_{1w} - R^+}{R^+ - R^-} + \Delta m_m(\tau_0)\frac{k_w}{R^+ - R^-} \tag{6}$$

The normalized deviation Δm_w of the longitudinal magnetization of water M_w(t) from the equilibrium value (M_w^{eq}) at the $\tau = \tau_0$ is presented [8] in eq.(7):

$$\Delta m_w(\tau_0) = e^{-\pi\omega_L\tau_0 - \left(\frac{\pi\omega_G\tau_0}{2(\ln 2)^{1/2}}\right)^2} - 1 \tag{7}$$

The water NMR signal (Fig. **1**) after FT was presented in the frequency domain as a spectral line with a shape best described by *a Voigt* function. The signal with the *Voigt* function is produced by a *Gaussian* broadening of the *Lorentzian* peak [8, 14, 17, 18]. ω_L and ω_G are the *Lorentzian* and *Gaussian* full widths at half-maximum height of the peaks in the frequency domain, respectively. It was experimentally found that at τ_0=65 μs the transverse magnetization of wood macromolecules $M_m(\tau_0)\approx 0$ in comparison with $M_w(\tau_0)$ and the normalized deviation $\Delta m_m(\tau_0)$ = $[M_m(\tau_0)$ – $M_m^{eq}]/M_m^{eq}$ from the equilibrium value of longitudinal magnetization of the macromolecule pool = –1 [8, 18].

The equations (2)-(7) can be fitted to the GS experimental results (*i.e.* according to Fig. **3**) yielding the estimates of k_w, R_{1w}, k_m and R_{1m} [8, 9, 17]. The authors of the work [20] evaluated the effect of cross-relaxation on the stimulated echo (STE) amplitude. Self-diffusion coefficient values were decreasing with increasing diffusion times. Final expression for echo-attenuation function displayed in the work [20] contains all the parameters of STE PFG experiment, including the diffusion constant, and a combination of four cross-relaxation parameters (k_w, R_{1w},

k_m and R_{1m}). When these parameters (k_w, R_{1w}, k_m and R_{1m}) are known, the diffusion coefficient can be determined by a nonlinear regression analysis fitting the echo-attenuation curve to derived expression [9, 20]. We have measured diffusion coefficient (D_{app}) in wood samples for two perpendicular directions (*X, Z*) applying the gradient of magnetic field with 0° or 90° to the static magnetic field B_0 and taking into account the cross-relaxation factor at analysis of the data obtained.

Fig. (**5**) compares the T_2 distributions (based on CPMG experiments) for wet wood (*top*: high level of moisture), and for wood with low level of moisture (*bottom*). The results testify that when the wood pieces become drier, the surface water is removed, and mobile water fractions with wide T_2 range between 40 and 150 ms (Fig. **5**) become less mobile (T_2= 40-80 ms) and less exchangeable with other compartments with water. Therefore, the peaks are narrow for much drier samples. The intensity of the most mobile T_2 peak decreased from ~60% to ~48%.

The study of drying process shows that when HL value in wood samples decreased from 1.303 to 0.525 g H_2O per g dry mass, the intensity of observable protons of water decreased from 100% to 34%. At HL=0.102, the component with T_2 in the range of 10-100 ms was practically absent. The peak with the bound water having T_2 around 1 ms was present in this sample. For this sample (HL=0.102), the cross-relaxation rate was defined from GS experiment (Fig. **3**) resulting in k_w=38.7 s^{-1}. For the wetter wood samples, *e.g.*, with HL=0.313 and HL = 0.556 g H_2O per g dry mass, the cross-relaxation rates were k_w=29.2 s^{-1} and 14.8 s^{-1}, respectively. These results testify that cross-relaxation was dependent on humidity of wood, and this is in line with preliminary published data showing that the value of k_w was dependent on HL value of the sample [18, 21]. The effect of the cross-relaxation on the longitudinal relaxation is significant, *i.e*, causing the bi-exponential behavior of the longitudinal magnetization recovery when the exchange rate of spin energy through the water-macromolecule interface exceeds the spin-lattice relaxation rate of bound water [17 - 19, 21].

The diffusion coefficient of bulk water was measured in doped water sample at the gradient in *X*- or *Z*-direction following the method outlined in chapter 1, eq. (6), *i.e.* $\ln(I(G)/I^*(0))$ was plotted against gradient factor $b=q^2(\Delta-\delta/3)$ for all values of diffusion time Δ. A linear fit was then made to the echo-attenuation data. The diffusion coefficients were then found from the plots and eq. (6) (chapter 1). The fits to the echo-attenuation data at all Δ values have been done for the *X*-gradient and for the *Z*-gradient. As expected, the diffusion coefficients for water did not depend on the diffusion time and direction of gradient. The values of $(2.33–2.42)\times10^{-9}$ m^2/s also agreed well with known published values for the bulk water diffusion coefficient, D_0 [22 - 27].

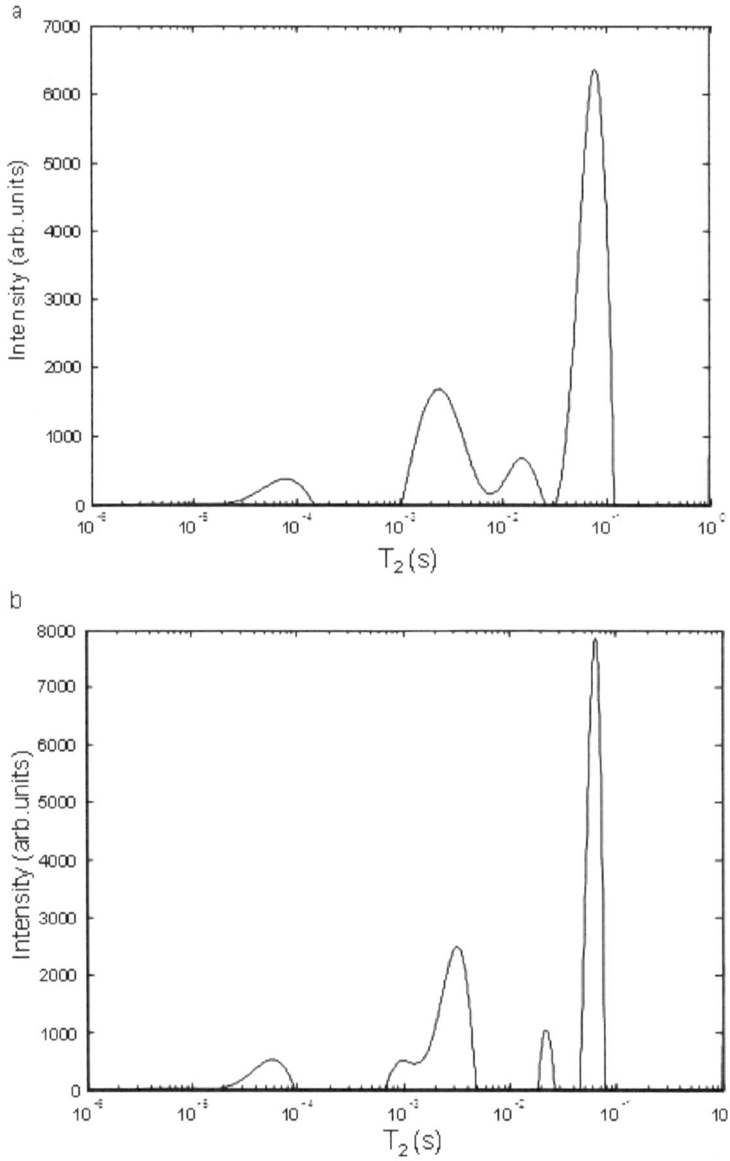

Fig. (5). T_2 distributions of ^{1}H obtained by 1D ILT treatment of CPMG data sets measured on the wood samples with HL=0.808 (**a**) and HL=0.525 (**b**) at 298 K and a frequency of 400 MHz. *X*-axis is presented in log-scale from 10^{-6} s to 1 s. The integral intensities of T_2 peaks in **a**-Graph (*from left to right*) are 5.5%, 26.8%, 7.5%, 60.2% of measurable protons. The integral intensities of T_2 peaks in the **b**-Graph (*from left to right*) are 8.3%, 38.9%, 4.6%, 48.1% of measurable protons.

When the gradient was oriented in radial/tangential direction to face of wood piece with HL=0.84 in PFG experiment at Δ= 20 ms, the value of apparent diffusion coefficient of water has been obtained as $D_{\Delta=20ms}=1.58\times10^{-9}$ m^2/s. CR effect discussed above has been taken into account at final calculation. There was no systematic variation in D values between the radial and tangential faces of wood pieces measured with the same experimental parameters. However, for the gradient aligned in wood fiber orientation (along static magnetic field B_0), the diffusion coefficient of water was $D_{\Delta=20ms}$ = (2.07-2.09) $\times10^{-9}$ m^2/s discovering anisotropic self-diffusion of water in the wood along the direction of the fibers and in perpendicular one.

At isotropic diffusion, all directions are equivalent. The wood is a fibrous material. Water diffusion in fibrous material is with preferred direction along the dominant fiber. In reality, the diffusion is influenced by a number of factors, in particular, pore size and pore size distribution, permeability of cell membranes and anisotropy [7 - 9, 13 - 15, 22]. The quantification of diffusion anisotropy in wood is very complex. Many factors can influence the rate of diffusion of water in wood. The barriers to diffusion are largely due to the nature and alignment of the tracheid cells. In natural environment of wood cells there is active transport of water along the length of the tracheid cell. However, there are barriers to transverse movement. As the diffusion rates are not equal in all directions the diffusion in wood is anisotropic. It may be related to structural morphology.

The tracheids found in spruce wood are mostly vertical narrow cells of length about ~2-3 mm with cell diameters about 20-40 μm. The characteristics of these cells determine the performances and efficiency of the xylem tissue within the tree. The cell walls of the tracheids cause the restrictions resulting in the water diffusion coefficient to be different along the different direction of diffusion. In spruce wood, water diffusion along the length of the cell is likely to be less restricted than diffusion in the transverse directions (radial and/ tangential). Thus, in 1D NMR diffusion experiment, the echo signal in the STE pulse sequence is dependent on the direction of the diffusion-encoding gradient relative to the wood cells. Diffusion coefficients obtained in 1D experiment on wood piece aligned along the static magnetic field B_0 were different at applying gradient in X- and Z-directions.

The $D_{x,z}$ values were also depending on diffusion time decreasing when Δ increased from 17-20 ms to 600-800 ms. In a number of publications [8, 9, 22 - 26], a dependence of $D_{app} = f(\Delta)$ is considered as a restriction phenomenon of water diffusion by barriers of macromolecular arrangement. Fig. (**6**) shows the dependences of diffusion coefficients in two perpendicular directions of applied gradients on the diffusion time Δ for the spruce wood. The experiments with gradient pulses in the transverse direction (X) showed variation of apparent diffusion coefficients D_{app} with Δ: from 1.62×10^{-9} m^2/s at Δ=17 ms to 0.24×10^{-9}

m^2/s at Δ=800 ms. The large decrease in D_{app} magnitude with increasing Δ (a factor of ~7) indicates the effects of barriers to diffusion. This is in line with the data of the publication [23] which presented the variation of the apparent diffusion coefficient of water in eastern white pine as a function of the diffusion time, Δ, indicating the effects of restriction to diffusion. D_{app} decreased with increasing Δ (a factor of 10) [23]. In contrast, the experiments with gradient pulses in the longitudinal directions in eastern white pine as a function of diffusion time showed the decrease in D_{app} at large Δ (about 17% at 903 ms). D_{app} was 2.0×10^{-9} m^2/s for the longest diffusion time of 903 ms whereas for Δ =13 ms D_{app} was 2.4×10^{-9} m^2/s [23]. The authors of this publication considered that the decrease in D_{app} at large Δ in the longitudinal direction may indicate a certain degree of restriction but not the impenetrable type as in tangential direction. A reduction in the diffusion coefficient with increased diffusion time for longitudinal direction was different for other type of wood and moisture level of material [23, 26]. If the diffusion coefficient is assumed to approach D_0 at short diffusion times, then a reduction of up to 50% in the sitca spruce wood (HL=0.70) was observed for D_{app} at Δ=200 ms [26]. There were no experiments at Δ>200 ms in [26]. The experiment at the shortest value of Δ=60 ms resulted in D_{app} $=1.75\times10^{-9}$ m^2/s which is lower than diffusion coefficient of free bulk water (about 26%) [26, 27]. According to the data obtained in PFG STE experiment (400 MHz) on the spruce wood at HL=0.84 for longitudinal direction (Fig. **6**), a reduction was from D_{app}=2.09×10^{-9} m^2/s (Δ=20 ms) to D_{app}=0.67×10^{-9} m^2/s (Δ=800 ms).

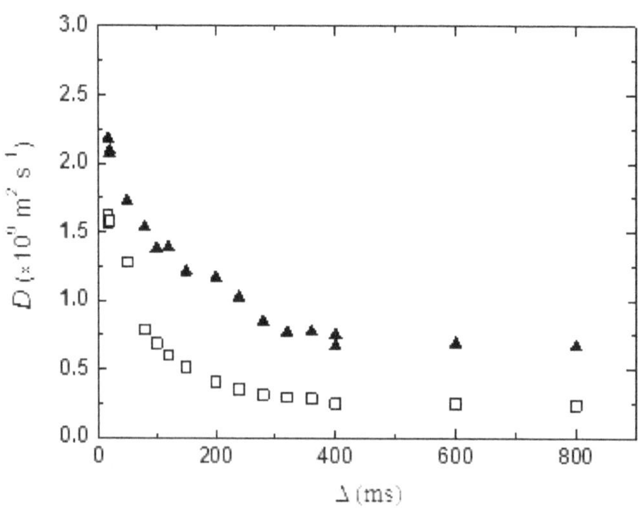

Fig. (6). The variation of the apparent diffusion coefficient in spruce wood as a function of the diffusion time, Δ, in the longitudinal (filled triangle) and in the tangential directions (opened squares). T=298 K, and frequency is 400 MHz. HL=0.84.

In the longitudinal orientation it would be expected that the diffusion of the water would experience much less restriction than in the radial or tangential orientations. This is due to the length of the tracheid cells which can be up to 100 times the width [23, 26, 28]. If to suppose that the diffusion is restricted (*i.e.* there are diffusion barriers) then the mean square displacement of the protons increases with less than the first power of the diffusion time ($<x^2> \sim \Delta^k$ with $k<1$) [8, 9, 14, 29] whereas apparent diffusion coefficient depends on Δ as $D_{app} \sim \Delta^{k-1}$. At completely restricted diffusion, $<x^2>$ does not increase with diffusion time, therefore $D_{app} \sim 1/\Delta$ (in accordance with $D_{app} = <x^2>/2\cdot\Delta$). In some works on biological porous systems, a reduction in the apparent diffusion coefficient on diffusion time are explained by a system of semipermeable barriers [8, 9, 22, 24, 25].

Methane and absorbed water have also used to probe the longitudinal wood cell structure of *Pinus sylvestris* [28, 30]. Longitudinal tracheids constitute 93% of the *Pinus sylvestris* volume. These works presented the pore size distribution measured in the longitudinal direction using methane as a probe. Two structural components were found in the fit of the diffusion data with mean lengths of 2.88 mm and 0.29 mm. These values were assigned to the longitudinal tracheid lumen and the height of the radial ray cells [28, 30].

It is considered that method PFG STE NMR with combined use of water and methane is applicable for investigating technological porous materials with pore dimensions ranging from 10 μm to 10 mm [28, 30]. It can be used also for studying the changes of the structure of wood at termal modifications [30, 31]. Thermal modification (*e.g.*, with varying temperatures between 180°C and 240°C) is a process that enhances properties of timber [31]. The T_2 relaxation time of water in the modified wood is longer than that in the nonmodified samples [30].

CONCLUDING REMARKS

Porous structure of wood and technological properties of timber can be studied by NMR relaxation and one-dimentional NMR diffusion of probe fluids. The most important fluid naturally existing in wood is water. So, the relaxation spectrum of water components and apparent diffusion coefficients measured in one-dimensional PFG NMR experiments on wood with different orientation of the samples to the magnetic field or different direction of gradients result in valuable information about the apparent size of cells in wood and anisotropic properties of water in timber. These approaches can potentially be used to diagnose and monitor the treatment of wood that involve macromolecular reorganization and associated changes in cross-relaxation and molecular diffusion of water.

The considered methods can be used in studying the decay in wood. The decay in wood is associated with the breakdown of cell walls causing the changes in the average size of cells [23], *i.e.*, increasing the distribution of sizes and resulting in water diffusion changes. Diffusion anisotropy could also provide information about the wood decaying.

ACKNOWLEDGEMENTS

The author thanks Marc Jones and Peter McDonald.

REFERENCES

[1] M. Lindner, W. Werhahn-Mees, T. Suominen, D. Vötter, S. Zudin, M. Pekkanen, R. Päivinen, M. Roubalova, P. Kneblik, F. Brüchert, E. Valinger, L. Guinard, and S. Pizzirani, "Conducting sustainability impact assessments of forestry – wood chains: Examples of ToSIA applications", *Eur. J. For. Res.*, vol. 131, no. 1, pp. 21-34, 2012. [http://dx.doi.org/10.1007/s10342-011-0483-7]

[2] W.T. Simpson, Drying and Control of Moisture Content and Dimensional Changes.*Forest Products Laboratory. Wood handbook –Wood as an engineering material.Gen. Tech. Rep. FPL–GTR–113.* U.S. Department of Agriculture, Forest Service, Forest Products Laboratory: Madison, WI, 1999.

[3] I. Santoni, E. Callone, A. Sandak, J. Sandak, and S. Dirè, "Solid state NMR and IR characterization of wood polymer structure in relation to tree provenance", *Carbohydr. Polym.*, vol. 117, pp. 710-721, 2015. [http://dx.doi.org/10.1016/j.carbpol.2014.10.057] [PMID: 25498692]

[4] L.G. Thygesen, and T. Elder, "Moisture in untreated, acetylated, and furfurylated norway spruce studied during drying using time domain NMR", *Wood Fiber Sci.*, vol. 40, pp. 309-320, 2008.

[5] V-Ve. Telkki, M. Yliniemi, and J. Jokisaari, "Moisture in softwoods: fiber saturation point, hydroxyl site content, and the amount of micropores as determined from NMR relaxation time distributions", *Holzforschung,* vol. 67, no. 3, pp. 291-300, 2013. [http://dx.doi.org/10.1515/hf-2012-0057]

[6] V.V. Rodin, P.J. McDonald, and M. Jones, "Two-dimensional distribution function of diffusion in wood obtained using 2D Laplace inversion", *Appl. Phys. Math.*, no. 6, pp. 3-7, 2014.

[7] C.A.S. Hill, A.J. Norton, and G. Newman, "The water vapour sorption properties of Sitka spruce determined using a dynamic vapour sorption apparatus", *Wood Sci. Technol.*, vol. 44, no. 3, pp. 497-514, 2010. [http://dx.doi.org/10.1007/s00226-010-0305-y]

[8] V.V. Rodin, "Methods of Magnetic Resonance in Studying Natural Biomaterials", In: *Encyclopedia of Physical Organic Chemistry,* Zerong Wang, Ed., Chapter 53. vol. 4, part 4. John Wiley & Sons, Inc., 2017, pp. 2861-2908.

[9] V.V. Rodin, and V.A. Nikerov, "NMR relaxation and PFG NMR studies of water dynamics in oriented collagen fibres with different degree of cross-linking", *Curr. Tissue Eng.*, vol. 3, no. 1, pp. 47-61, 2014. [http://dx.doi.org/10.2174/2211542003666140626211652]

[10] P. Fantazzini, A. Maccotta, M. Gombia, C. Garavaglia, R.J. Brown, and M. Brai, "Solid–liquid nuclear magnetic resonance relaxation and signal amplitude relationships with ranking of seasoned softwoods and hardwoods", *J. Appl. Phys.*, pp. 0749071-0749077, 2006.

[11] V.V. Rodin, P.J. McDonald, and S. Zamani, "A nuclear magnetic resonance pulsed field gradient study of self-diffusion of water in hydrated cement pastes", *Diffus. Fundam,* vol. 18, pp. 1-7, 2013.

[12] A. Valori, V.V. Rodin, and P.J. McDonald, "On the interpretation of ^{1}H 2-dimensional NMR relaxation exchange spectra in cements: is there exchange between pores with two characteristic sizes or Fe^{3+} concentrations?", *Cement Concr. Res.*, vol. 40, no. 9, pp. 1375-1377, 2010. [http://dx.doi.org/10.1016/j.cemconres.2010.03.022]

[13] P.T. Callaghan, *Translational Dynamics and Magnetic Resonance.* Oxford University Press: Oxford, 2011. [http://dx.doi.org/10.1093/acprof:oso/9780199556984.001.0001]

[14] V.V. Rodin, *Magnetic Resonance Methods.* Press MIPT: Moscow, 2004.

[15] P.T. Callaghan, "How two pairs of gradient pulses give access to new information about molecular dynamics", *Diffus. Fundam.*, vol. 2, no. 64, pp. 1-18, 2005.

[16] J. Cox, P.J. McDonald, and B.A. Gardiner, "A study of water exchange in wood by means of 2D NMR relaxation correlation and exchange", *Holzforschung,* vol. 64, no. 2, pp. 259-266, 2010. [http://dx.doi.org/10.1515/hf.2010.036]

[17] V.V. Rodin, L. Foucat, and J-P. Renou, "Natural polymers according to NMR data: cross-relaxation in collagen macromolecules from two samples of connective tissues", *Biophysics,* vol. 49, no. 4, pp. 563-571, 2004. [PMID: 15458243]

[18] V.V. Rodin, and D.P. Knight, "Natural materials by NMR data: cross-relaxation in macromolecules of natural silk", *Mater. Sci.,* no. 10, pp. 16-21, 2003.

[19] V.V. Rodin, "NMR Relaxation in investigation of water in hydrated collagen of Lesser Spotted Dogfish", *Biophysics,* vol. 50, no. 2, pp. 219-225, 2005.

[20] L.J. Peschier, J.A. Bouwstra, J. de Bleyser, H.E. Junginger, and J.C. Leyte, "Cross-relaxation effects in pulsed-field-gradient stimulated-echo measurements on water in a macromolecular matrix", *J. Magn. Reson. B.,* vol. 110, no. 2, pp. 150-157, 1996. [http://dx.doi.org/10.1006/jmrb.1996.0024]

[21] M-C. Vackier, B.P. Hills, and D.N. Rutledge, "An NMR relaxation study of the state of water in gelatin gels", *J. Magn. Reson.,* vol. 138, no. 1, pp. 36-42, 1999. [http://dx.doi.org/10.1006/jmre.1999.1730] [PMID: 10329223]

[22] J.E. Tanner, and E.O. Stejskal, "Restricted self-diffusion of protons in colloidal systems by the pulsed-gradient spin-echo method", *J. Chem. Phys.,* vol. 49, no. 4, pp. 1768-1777, 1968. [http://dx.doi.org/10.1063/1.1670306]

[23] W. Wycoff, S. Pickup, B. Cuter, B. Cutter, W. Miller, and T.C. Wong, "The determination of the cell size in wood by nuclear magnetic resonance diffusion techniques", *Wood Fiber Sci.,* vol. 32, no. 1, pp. 72-80, 2000.

[24] J.E. Tanner, "Self diffusion of water in frog muscle", *Biophys. J.,* vol. 28, no. 1, pp. 107-116, 1979. [http://dx.doi.org/10.1016/S0006-3495(79)85162-0] [PMID: 318065]

[25] A. Traore, *Etude par RMM et diffusion des neutrons de la dynamique de l'eau dans le collagene nativ et dénaturé. These.* Universite Blaise Pascal, 1997.

[26] M. Jones, PhD Thesis: University of Surrey, Guildford, 2013.

[27] V.V. Rodin, and T. Cosgrove, "Nuclear magnetic resonance study of water-polymer interactions and self-diffusion of water in polymer films", *OALJ: Chem. Mater. Sci.,* vol. 3, no. 10, pp. 1-17, 2016.

[28] P.M. Kekkonen, V-V. Telkki, and J. Jokisaari, "Determining the highly anisotropic cell structures of Pinus sylvestris in three orthogonal directions by PGSTE NMR of absorbed water and methane", *J. Phys. Chem. B,* vol. 113, no. 4, pp. 1080-1084, 2009. [http://dx.doi.org/10.1021/jp807848d] [PMID: 19105729]

[29] R. Trampel, J. Schiller, L. Naji, F. Stallmach, J. Kärger, and K. Arnold, "Self-diffusion of polymers in

cartilage as studied by pulsed field gradient NMR", *Biophys. Chem.,* vol. 97, no. 2-3, pp. 251-260, 2002.
[http://dx.doi.org/10.1016/S0301-4622(02)00078-9] [PMID: 12050014]

[30] P. Kekkonen, *"Characterization of thermally modified wood by nmr spectroscopy: microstructure and moisture components"*, PhD Thesis, University of Oulu, Oulu, 2014.

[31] M.A. Javed, P.M. Kekkonen, S. Ahola, and V-V. Telkki, "Magnetic resonance imaging study of water absorption in thermally modified pine wood", *Holzforschung,* vol. 69, no. 7, pp. 899-907, 2015.
[http://dx.doi.org/10.1515/hf-2014-0183]

CHAPTER 6

2D Diffusion-Diffusion Correlation NMR Spectroscopy in Study of Diffusion Anisotropy in Wood

Abstract: This chapter describes how to use 2-dimensional pulsed field gradient (PFG) NMR spectroscopy to explore the anisotropic diffusion of water in wood. Simulations based on the theory of 2D diffusion-diffusion correlations with the application of inverse Laplace transform (ILT) have been carried out for orthogonal and collinear pairs of gradients with parameters of 2D experiment on wood. The algorithm uses analytical solution for the echo attenuation in the 2D PFG experiment with collinear/orthogonal applications of two gradient pulses pairs. Numerical expressions calculated for the gradients and the model diffusion coefficients in both orthogonal directions were treated by 2D ILT on various artificial and experimental data sets. The algorithm is presented in a NMR context for wetting wood when the gradient directions are switched. The results showed how 2D (D_1, D_2) ILT maps reflected different ratios D_1/D_2 in two orthogonal directions. When two pairs of gradients were applied orthogonally in 2D experiment on wood, the features of diffusion anisotropy appeared as off-diagonal "wings" in 2D diffusion map whereas in the experiment with two collinear pairs of gradients, anisotropic diffusion of water in wood cells has been observed as two diagonal peaks in 2D map.

Keywords: 2D diffusion-diffusion correlation NMR spectroscopy (DDCOSY), Anisotropic diffusion, Apparent diffusion coefficient D_{app}, Free induction decay (FID), Inverse Laplace transform (ILT), NMR, Pulsed Field Gradient (PFG) NMR, Spin-echo (SE), Spin-lattice (longitudinal) relaxation time T_1, Spin-spin (transverse) relaxation time T_2, Stimulated echo (STE), Water, Wood.

INTRODUCTION

NMR diffusion methods (PFG spin-echo (SE) and PFG stimulated echo (STE)) are used in the study of fluids in such heterogeneous materials as wood [1 - 3]. Chapter 5 showed the possibilities of NMR relaxation and one-dimensional NMR diffusion in studying water compartments in wood and anisotropic water diffusion in wood cells [2 - 6].

Victor V. Rodin

In addition to 1D NMR methods, 2D NMR techniques which produce second order data matrices are also applied [7 - 14]. 2D NMR approaches were frequently used in the characterization of relaxation times in porous media, for example, hydrating cement pastes [11, 12]. The fitting of a sum of exponentials to the measured 2D response data is a main procedure of these NMR techniques [7 - 9, 11 - 14]. Many model studies develop such a fitting procedure with the aid of multidimensional inverse Laplace transform (ILT) [15 - 22].

2D NMR diffusion-diffusion correlation spectroscopy (DDCOSY) has been applied to study anisotropic properties of such materials as liquid crystals, plant tissues, cement, collagen fibers, wood [7 - 9, 11 - 14, 22 - 24]. These models could be developed for studying the wetting / drying phenomena in some fibrous materials with anisotropic mobility of water [7, 8].

This chapter analyzes wood-water interactions and anisotropic diffusion of water in spruce wood using ^1H NMR pulsed field gradient (PFG) diffusometry in two-dimensions. The paper shows DDCOSY results on oriented wood samples applying two pairs of gradients in orthogonal/collinear directions. Simulations based on the theory of 2D diffusion-diffusion correlations with following application of ILT have been carried out for orthogonal and collinear pairs of gradients studying the models of isotropic and anisotropic diffusion with parameters of 2-dimensional experiment on wood. 2D NMR DDCOSY experiments on wood with two pairs of orthogonal gradients result in 2D diffusion map with off-diagonal "wings" reflecting diffusion anisotropy. For 2D NMR DDCOSY experiment with two collinear pairs of gradients, anisotropic diffusion of water in wood has been observed as 2D diffusion map with two diagonal spots.

THEORY AND MODELS IN DIFFUSION-DIFFUSION CORRELATION SPECTROSCOPY (DDCOSY)

Two-dimensional DDCOSY experiments comprising of two spin-echo (SE) pulse sequences with two pairs of gradients were carried out with encoding times of 6 ms and gradients in the *X*-direction (Fig. **7**, *bottom,* Chapter 1) or in both *X*- and *Y*-directions [7 - 9, 11]. Echo decays were inverted using the 2D ILT algorithm according to Venkataramanan *et al.* [19, 20]. Further the chapter presents additional details of the ILT algorithm at modeling diffusion anisotropy.

Bulk water sample was measured in DDCOSY NMR experiments also to test the pulse sequences for pairs of collinear (and orthogonal) gradients and clarify experimental parameters for the studies of wood [8, 9].

In common two-dimensional NMR spectroscopy, the signal is measured as function of two independent variables, *e.g.*, such as times, giving a possibility the spin system to evolve at different mechanisms. These two-dimensional data are managed with the help of the efficient algorithms, such as fast Fourier transform [14]. To study a dynamics of molecular species, *e.g.*, on the base of relaxation properties, the times of spin-lattice and spin-spin relaxation (T_1 and T_2) may be used. Two-dimensional correlation functions for these relaxation constants can be obtained in the experiments which are analogous to those in 2D NMR spectroscopy. The analysis of these correlation data requires multidimensional ILT [13 - 16]. In the works [19, 20] on the example of T_1-T_2 correlation experiment, the authors did analysis of the acquired data as a two-dimensional array and related the signal to the probability density of T_1 and T_2 *via* integral, an example of a broad class of Fredholm integrals of the first kind. Those integrals are frequently considered in different applications [21]. The authors of the work [19] emphasized that the solution for Fredholm integrals is often considered with certain assumptions about the character of experimental noise and probability density of the molecules with relaxation times of T_1 and T_2. Although there are several algorithms for 1D ILT, the extension of these algorithms to 2D makes enormous difficulties on computer memory. In the work [20], the authors described an effective algorithm to solve this problem and implemented this on the examples of T_1-T_2 correlation experiments [19].

In order to simulate two-dimensional diffusion experiments on wood, we describe the diffusion anisotropy with the aid of same scheme that was applied in the works with 2D DDCOSY experiments (Fig. **7**, *bottom,* Chapter 1) on different materials [7 - 9, 13, 22]. For implementation of algorithm, the scheme with cylindrical symmetry was considered presenting the polar geometry of randomly oriented locally anisotropic diffusion domains [22, 24]. Diffusion anisotropy is considered in molecular frame associated with tracheid cells [7, 8]. Similar to the theory of 2D diffusion-diffusion correlations [8, 13, 14, 22] in a local domain, the diffusion tensor in wood is presented by axial symmetry and it is characterized by two diffusivities D_1 and D_2 which are parallel to and perpendicular to the local principal axis in molecular frame. With a transformation giving by the inverse Laplace, it would be possible to calculate the distribution of apparent diffusion tensor elements D_1, D_2 in wood [7, 8].

The details of the mathematical model were developed on T_1-T_2 correlations [19]. The model was applied to the distribution of diffusion tensor elements D_1, D_2 in 2D SE NMR study [7, 11, 13, 22]. The data $I\,(q_1^2,q_2^2)$ for attenuation of spin-echo signal with respect to the independent q_1^2 and q_2^2 variables connects to the kernel function k_1k_2 (which in the case of exponential decays for both orthogonal directions 1 and 2 is presented as $\exp(-q_1^2D_1\Delta)\exp(-q_2^2D_1\Delta)$) by the Fredholm integral of the first kind in eq. (1):

$$I(q_1^2,q_2^2)=\iint k_1(D_1,q_1^2)k_2(D_2,q_2^2)f(D_1,D_2)dD_1dD_2+R_N(q_1^2,q_2^2) \quad \textbf{(1)}$$

Here $R_N(q_1^2,q_2^2)$ is the experimental noise. It is assumed to be additive white Gaussian noise. The function $f(D_1, D_2)$ corresponds to the probability density of molecules with diffusion coefficients of D_1, D_2. $f(D_1, D_2) \geq 0$ for all D_1 and D_2. By analogy with 2D NMR spectroscopy, $f(D_1, D_2)$ is regarded as 2D spectrum of D_1, D_2 or 2D map. The variables $q_1=\gamma\delta_1 G_1$ and $q_2=\gamma\delta_2 G_2$ are the wave vectors for two pairs of gradient pulses with length δ_1 and δ_2 oriented along *X/Y/Z*-directions. The parameters and directions of the 1st and 2nd gradients can be changed independently.

According to the mathematical model of T_1-T_2 correlations [19] the eq. (1) can be approximated by the matrix of the discretized form and be re-written into the eq. (2):

$$I=K_1FK_2^{'}+R_N \quad \textbf{(2)}$$

where the matrices K_1, K_2 and F denote discretized versions of k_1, k_2 and f, respectively. These matrices have the dimensions $N_1 \times N_x$, $N_2 \times N_y$, and $N_x \times N_y$, respectively [19].

Venkataramanan and Song in the work [20] solved the class of the 2D Fredholm integrals which for the case of echo-attenuation in 2D diffusion experiment can be presented here by eq. (2). The algorithm of Venkataramanan and Song [19, 20] allows one to perform two-dimensional numerical inverse Laplace transformation. The algorithm solves the double integral equation for signal of echo as a function of two variables q_1 and q_2. Therefore, one needs to transfer echo-attenuation function (with tensor D_1 and D_2 elements in molecular frame) into laboratory frame through the consequent rotations [8, 13, 22, 24]. In this DDCOSY experiment (Fig. **7**, *bottom,* Chapter 1) the local diffusion anisotropy can be probed by changing the relative directions of q_1 and q_2. As a result, the total echo signal in 2D DDCOSY experiment could be considered in two-dimensional space and this is obtained from linear superposition of q_1, q_2 in both directions [7, 13, 22].

With the approach described above we are looking for anisotropy and local order in molecular frame whereas NMR diffusion experiment is carried out in laboratory frame *XYZ*. We consider randomly oriented molecular frames. In this case, taking into account the summary over all possible orientation of local directors of molecular coordinate system, the resulting signals in 2D space of laboratory frame for the cases of orthogonal and collinear gradients are given in

Refs. [13, 14, 22, 25]. On the base of those equation, the matrices of echo signals can be numerically calculated and be transformed into 2D spectra using 2D ILT.

DDCOSY WITH PARALLEL ORIENTED PAIRS OF GRADIENTS. SIMULATIONS AND EXPERIMENTS ON WOOD

In accordance with the theory of 2D diffusion-diffusion correlations [13, 22, 24], we consider a local domain in which the diffusion tensor has axial symmetry. It is characterized by two diffusivities D_1 and D_2 to describe parallel and perpendicular movement to the local principal axis in molecular frame, respectively. The matrices $I(q_1^2,q_2^2)$ for different values of D_1 and D_2 have been calculated using parameters from experiments on wood (gradient pulse length, diffusion times and values of gradients). The 2D DDCOSY experiments gave the parameters to be introduced into q_1 and q_2 values and produce the matrices of the echo signals. Further, the signal analysis used two-dimensional ILT to produce the response in 2D map for diffusion tensor elements.

Using the algorithm of Venkataramanan and Song [19, 20] in solving the double integral equation for signal of echo, the simulations have been done for the wave vectors q_1 and q_2 oriented in the same directions (the case of collinear gradients). Fig. (**1**) shows raw data plotted from numerically calculated 2D arrays and final 2D diffusion-diffusion correlation maps after 2D ILT for the case of anisotropic diffusion for D_2=5D_1=10^{-9} m^2/s. For comparison, the 2D map of isotropic diffusion with D_1=D_2=10^{-9} m^2/s is also presented. In first case, two diagonal peaks are reflecting diffusion anisotropy with different diffusion constants D_1 and D_2. In second case, 2D map of isotropic diffusion with D_1=D_2=10^{-9} m^2/s is shown by one round spot on the diagonal.

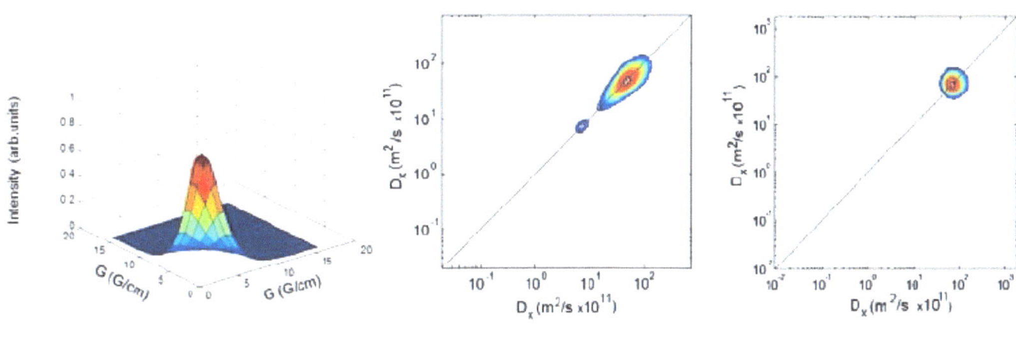

Fig. (1). Simulated 2D raw data (*left*, D_2=5D_1) and 2D (D_1,D_2) spectra at modeling diffusion experiment with the collinear pairs of magnetic field gradient pulses. D_2=5D_1=10^{-9} m^2/s (*centre*). D_2=D_1=10^{-9} m^2/s (*right*); Diffusion anisotropy is characterised by two extended spots on the diagonal (*centre*). Isotropic diffusion is highlighted as one round spot on the diagonal (*right*).

Fig. (**2**) shows 2D map as distribution of diffusion coefficients D_1 and D_2 in wood sample obtained in the 2D DDCOSY experiment with two pairs of collinear gradients. The data results in two peaks on the diagonal of 2D diffusion-diffusion correlation map. Thus, this 2D spectrum reflects diffusion anisotropy of water in wood cells with different D_1 and D_2 diffusion coefficients as elements of diffusion tensor.

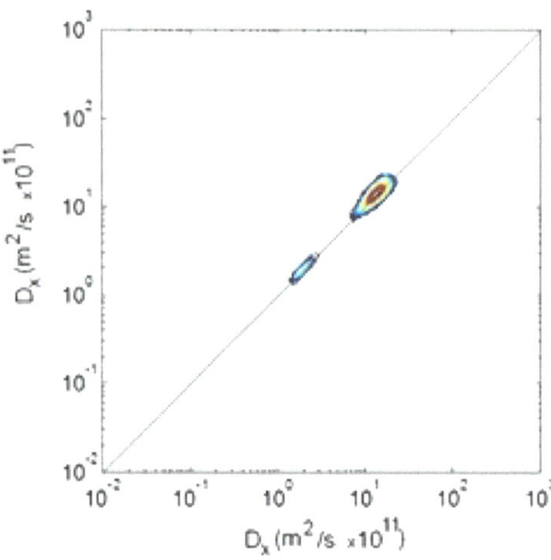

Fig. (2). 2D (D_1, D_2) spectra obtained from the data set of 2D DDCOSY experiment on wood sample with HL=61% carried out for collinear pairs of gradient pulses. Diffusion anisotropy is characterised by two extended spots on the diagonal.

These results presenting two separate spots on the diagonal of 2D correlation map (Fig. **2**) were also in agreement with simulation studies [8] as shown in Fig. (**1**). Similar 2D DDCOSY experiment has been carried out on pure water also [11]. These data resulted in one diagonal peak [7, 11] that was in agreement with simulated results for isotropic diffusion (Fig. **1**, *right*).

2D DDCOSY WITH PERPENDICULAR PAIRS OF GRADIENTS. INVERSE LAPLACE TRANSFORM IN STUDYING DIFFUSION ANISOTROPY IN WOOD

2D (D_1, D_2) spectra have been calculated for isotropic distribution of locally anisotropic diffusion elements. Matrix data sets were obtained using *MatCad* numerical integration of the equations for the echo-attenuation [13, 22, 24]. The matrices $I(q_1^2, q_2^2)$ for different values of D_1 and D_2 have been calculated with parameters (gradient pulse length, diffusion times and data set for values of gradients) taken from experiments on wood samples [7]. The 2D DDCOSY experiments gave the parameters to be introduced into q_1 and q_2 values and produced the matrices of the signals. Furthermore, the analysis of signal used two-dimensional ILT to calculate the response in 2D distribution of diffusion coefficients. Fig. (**3**) shows the resulting simulated 2D spectra for orthogonal gradient pulse pairs in the cases of the isotropic diffusion with $D_1 = D_2$ and anisotropic diffusion with $D_2 = 10D_1$. In last case, anisotropic structure is presented on the molecular level (in the molecular frame [8, 13]).

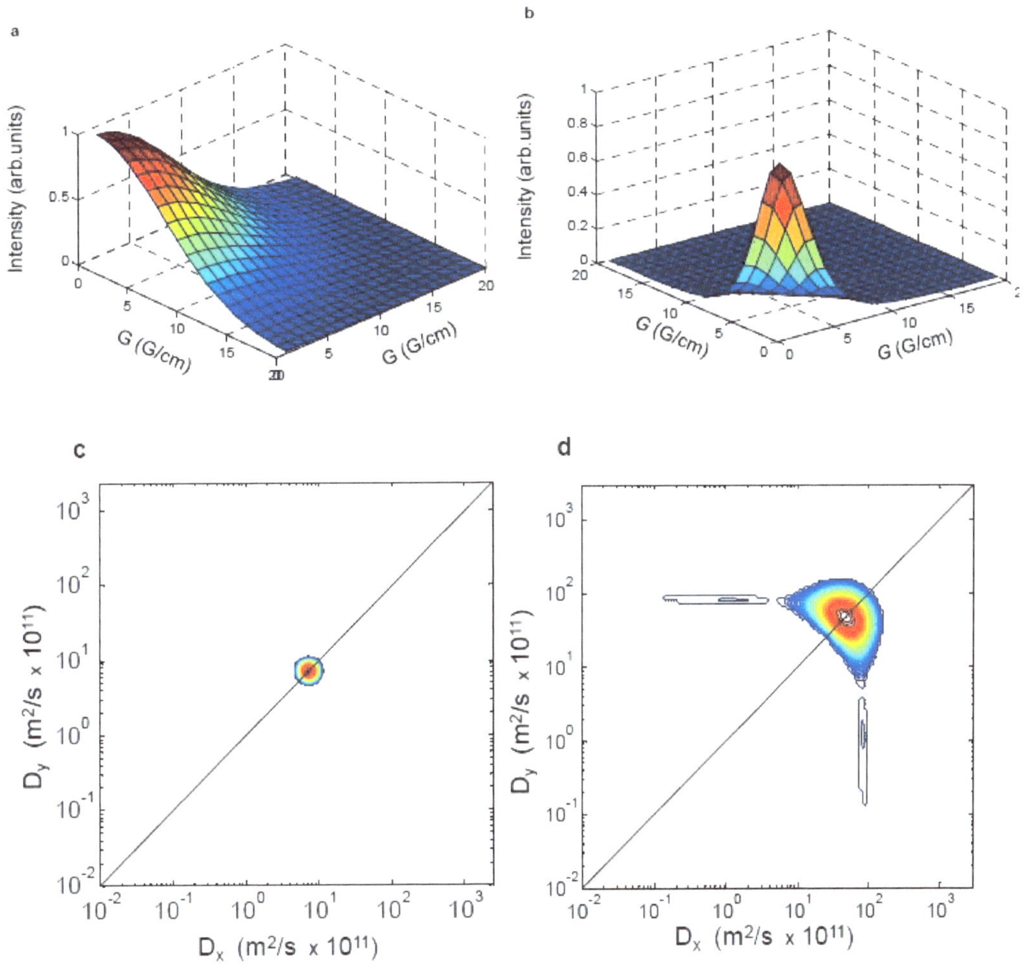

Fig. (3). Simulated 2D raw data (**a**, D_2=D_1 and **b**, D_2=10D_1) and 2D (D_1,D_2) spectra (**c**, **d**) at modeling diffusion experiment with the orthogonal pairs of magnetic field gradient pulses. D_2=D_1=10^{-10} m^2/s (**c**). D_2=10D_1=10^{-9} m^2/s (**d**); Diffusion anisotropy is characterised by two off-diagonal "wings" and main spot on the diagonal (**d**). Isotropic diffusion is highlighted as one round spot on the diagonal (**c**).

This reflects different diffusion constants D_1 and D_2. These simulations showed strong off-diagonal features (called as "long spots" or "wings"). These characteristics are the result of directly calculated spectra by the two-dimensional ILT. Fig. (**4**) shows the resulting 2D spectra for the DDCOSY experiments on wood with two pairs of orthogonal gradients. Similar 2D DDCOSY experiment with two pairs of orthogonal gradients has been performed for pure water: the result was observed as one diagonal peak (Fig. **5**) similar to the result which was simulated for isotropic diffusion (Fig. **3c**).

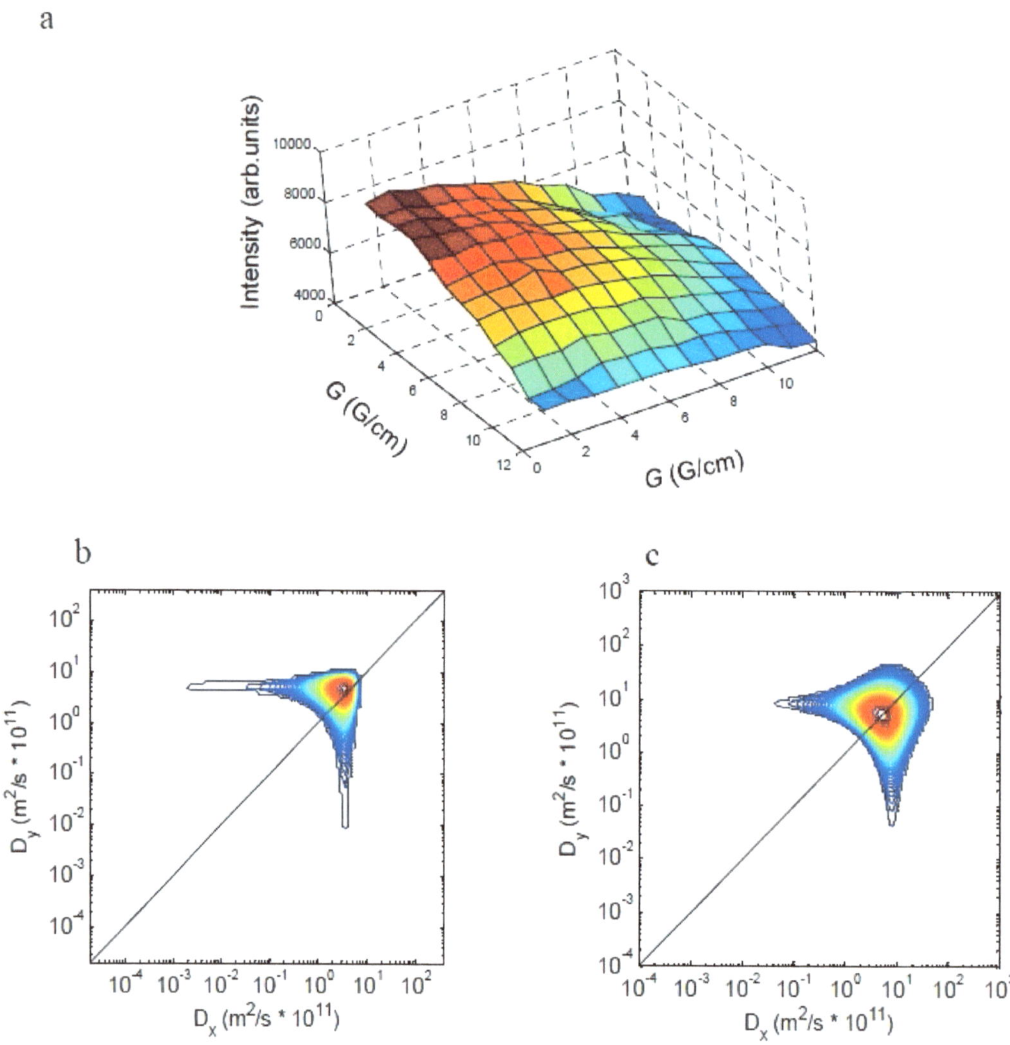

Fig. (4). 2D raw data from 2D DDCOSY experiment on wood (HL=51%, **a**) and 2D (D_1, D_2) spectra obtained on wood samples with HL=51% (**b**) and HL=59% (**c**) for orthogonal pairs of gradient pulses. Diffusion anisotropy is characterised by off-diagonal "wings".

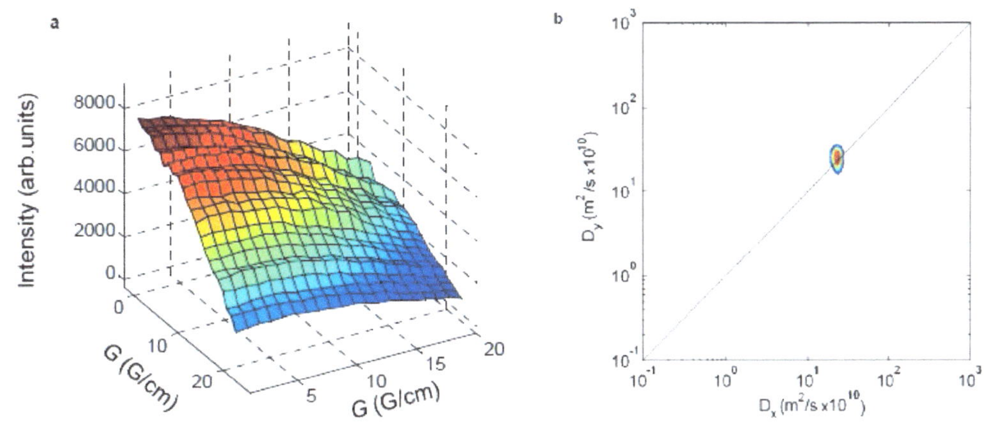

Fig. (5). 2D raw data from 2D DDCOSY experiment on pure water (**a**) and 2D (D_1, D_2) map obtained on pure water for orthogonal pairs of gradient pulses (**b**). Isotropic diffusion is characterised by one spot on the diagonal.

Thus, anisotropic diffusion of water in wood is characterized in DDCOSY experiment with orthogonal pairs of gradient pulses as "wings" outside main diagonal peak in 2D spectrum. Fig. (**4**) also compares the results on wood samples with different water content. The 2D map from wood with higher water content has clear differences around the main intensive peak on the diagonal and illustrates that the water diffuses faster along the plane of the wood cell than orthogonal to the tracheids. These data also testifies that apparent diffusion coefficient of water (longitudinal direction) in wet wood (HL=59%) is about 2-3 times larger than that in the wood sample with smaller humidity level (HL=51%). The 2D DDCOSY experiment is potentially able to monitor the development of changes in moisture level of wood materials measuring the intensity and position of main diagonal peak in 2D diffusion-diffusion correlation map keeping the correlation between D_1 and D_2 as "wings" and comparing these for wet wood and for wood pieces with low water content.The limits of methodology could be observed when HL level in the wood samples is close to FSP. Other limitation can be associated with high value of HL in wood, *i.e.*, when all the spaces between the cells are with water completely and the wood cells are swollen. In such a case, ILT can produce 2D diffusion-diffusion correlation map in the kind of one big intensive spot on the diagonal without any feature of "wings" outside diagonal as mirror of anisotropic diffusion. It is also known that the extent of water diffusion anisotropy in fibrous natural polymers depends on HL values [9, 25 - 29]. Therefore, 2D ILT DDCOSY spectra of wetting wood seem to very important because those would produce additional information to T_2 maps of absorbed water in wood materials. For example, it was shown [30] that T_2 maps of absorbed water in pine wood based on MRI study are not enough to characterise correct lumen size in thermally modified pine wood, and NMR diffusion study should be definitely involved to make right conclusion.

CONCLUSION

It has been successfully reported how 2D DDCOSY nuclear magnetic resonance methodology works in studies of anisotropic diffusion of water in wood. Diffusion-diffusion correlation spectroscopy has been used to present the results of PFG studies on wood. 2D diffusion-diffusion correlation map characterizes features of anisotropic diffusion of water as "wings" to main diagonal peak for the 2D PFG experiment with two pairs of orthogonal gradient pulses, and as two diagonal peaks for 2D PFG experiment with two pair of collinear gradients. The study compared wood pieces at different moisture content. The results showed clearly how the features of water distribution in anisotropic environment within the wood sample changed between the higher moisture content of wood piece and the lower moisture level of wood sample.

Wetting the wood to very high moisture content can result in the disappearance of the "wings" registered in 2D diffusion maps obtained by ILT from 2D DDCOSY experiments with orthogonal pairs of gradient pulses. Simulations of two-dimensional NMR experiments have been done for explanation of 2D results on the anisotropy of diffusion coefficients. The correlation of the diffusion motion of molecules along either collinear or orthogonal directions of applied pulsed gradients of magnetic field has been shown on the wood samples.

These methods are applicable for investigating wetting/drying wood, and it can be used for determining how processes of engineering and modifications in wood technology affect the diffusion anisotropy of water in wood cells. The use of 2D Laplace inversion algorithm in fast and efficient production of DDCOSY correlation spectra enables one to use two-dimensional diffusion-diffusion correlation spectroscopy for the study of molecular dynamics at collinear/orthogonal pairs of gradient. The combination of NMR relaxation and diffusion (one- and two-dimensional) techniques as well as appropriate model of the processes/phenomena is effective experimental methodology which is able to produce valuable information on the anisotropy and dynamic properties. This chapter considered the additional material which would allow one to get more details in description of diffusion anisotropy and properties of natural materials for technological applications in future.The processes associated with modification of wood can result in changes of diffusion anisotropy. This is also essential for the studies of the decaying wood.

ACKNOWLEDGEMENTS

The author thanks Marc Jones and Peter McDonald.

REFERENCES

[1] P.M. Kekkonen, V-V. Telkki, and J. Jokisaari, "Determining the highly anisotropic cell structures of Pinus sylvestris in three orthogonal directions by PGSTE NMR of absorbed water and methane", *J. Phys. Chem. B,* vol. 113, no. 4, pp. 1080-1084, 2009. [http://dx.doi.org/10.1021/jp807848d] [PMID: 19105729]

[2] P. Kekkonen, *"Characterization of thermally modified wood by nmr spectroscopy: microstructure and moisture components"*, PhD Thesis, University of Oulu, Oulu, 2014.

[3] E.H. Johannessen, E.W. Hansen, and J.B. Rosenholm, "Fluid self-diffusion in Scots pine sapwood tracheid cells", *J. Phys. Chem. B,* vol. 110, no. 5, pp. 2427-2434, 2006. [http://dx.doi.org/10.1021/jp0556431] [PMID: 16471834]

[4] L.G. Thygesen, and T. Elder, "Moisture in untreated, acetylated, and furfurylated norway spruce studied during drying using time domain NMR", *Wood Fiber Sci.,* vol. 40, pp. 309-320, 2008.

[5] V-Ve. Telkki, M. Yliniemi, and Ju. Jokisaari, "Moisture in softwoods: fiber saturation point, hydroxyl site content, and the amount of micropores as determined from NMR relaxation time distributions", *Holzforschung,* vol. 67, no. 3, pp. 291-300, 2013. [http://dx.doi.org/10.1515/hf-2012-0057]

[6] T.M. Todoruk, I.D. Hartley, R. Teymoori, J. Liang, and H. Peemoeller, " ^{1}H Nuclear magnetic resonance of Lodgepole pine wood chips affected by the mountain pine beetle", *Materials (Basel),* vol. 4, no. 1, pp. 131-140, 2010. [http://dx.doi.org/10.3390/ma4010131] [PMID: 28879981]

[7] V.V. Rodin, P.J. McDonald, and M. Jones, "Two-dimensional distribution function of diffusion in wood obtained using 2D Laplace inversion", *Appl. Phys. Math.,* no. 6, pp. 3-7, 2014.

[8] V.V. Rodin, "Methods of Magnetic Resonance in Studying Natural Biomaterials", In: *Encyclopedia of Physical Organic Chemistry,* Zerong Wang, Ed., Chapter 53. vol. 4, part 4. John Wiley & Sons, Inc., 2017, pp. 2861-2908.

[9] V.V. Rodin, and V.A. Nikerov, "NMR relaxation and PFG NMR studies of water dynamics in oriented collagen fibres with different degree of cross-linking", *Curr. Tissue Eng,* vol. 3, no. 1, pp. 47-61, 2014. [http://dx.doi.org/10.2174/2211542003666140626211652]

[10] P. Fantazzini, A. Maccotta, M. Gombia, C. Garavaglia, R.J.S. Brown, and M. Brai, "Solid–liquid nuclear magnetic resonance relaxation and signal amplitude relationships with ranking of seasoned softwoods and hardwoods", *J. Appl. Phys.,* vol. 100, pp. 0749071-0749077, 2006.

[11] V.V. Rodin, P.J. McDonald, and S. Zamani, "A nuclear magnetic resonance pulsed field gradient study of self-diffusion of water in hydrated cement pastes", *Diffus.Fundam,* vol. 18, no. 3, pp. 1-7, 2013.

[12] A. Valori, V.V. Rodin, and P.J. McDonald, "On the interpretation of ^{1}H 2-dimensional NMR relaxation exchange spectra in cements: is there exchange between pores with two characteristic sizes or Fe^{3+} concentrations?", *Cement Concr. Res.,* vol. 40, no. 9, pp. 1375-1377, 2010. [http://dx.doi.org/10.1016/j.cemconres.2010.03.022]

[13] P.T. Callaghan, "How two pairs of gradient pulses give access to new information about molecular dynamics", *Diffus. Fundam.,* vol. 2, no. 64, pp. 1-18, 2005.

[14] P.T. Callaghan, *Translational Dynamics and Magnetic Resonance.* Oxford University Press: Oxford, 2011. [http://dx.doi.org/10.1093/acprof:oso/9780199556984.001.0001]

[15] J.B. Moody, and Y. Xia, "Analysis of multi-exponential relaxation data with very short components using linear regularization", *J. Magn. Reson.,* vol. 167, no. 1, pp. 36-41, 2004. [http://dx.doi.org/10.1016/j.jmr.2003.11.004] [PMID: 14987596]

[16] P. Berman, O. Levi, Y. Parmet, M. Saunders, and Z. Wiesman, "Laplace inversion of low-resolution NMR relaxometry data using sparse representation methods", *Concepts Magn. Reson., Part A,* vol. 42, no. 3, pp. 72-88, 2013. [http://dx.doi.org/10.1002/cmr.a.21263] [PMID: 23847452]

[17] J. Granwehr, and P.J. Roberts, "Inverse Laplace transform of multidimensional relaxation data without non-negativity constraint", *J. Chem. Theory Comput.,* vol. 8, no. 10, pp. 3473-3482, 2012. [http://dx.doi.org/10.1021/ct3001393] [PMID: 26592997]

[18] S. Ahola, V.V. Zhivonitko, O. Mankinen, G. Zhang, A.M. Kantola, H-Y. Chen, C. Hilty, I.V. Koptyug, and V.V. Telkki, "Ultrafast multidimensional Laplace NMR for a rapid and sensitive chemical analysis", *Nat. Commun.,* vol. 6, p. 8363, 2015. [http://dx.doi.org/10.1038/ncomms9363] [PMID: 26381101]

[19] Y.Q. Song, L. Venkataramanan, M.D. Hürlimann, M. Flaum, P. Frulla, and C. Straley, "T_1-T_2 correlation spectra obtained using a fast two-dimensional Laplace inversion", *J. Magn. Reson.,* vol. 154, no. 2, pp. 261-268, 2002. [http://dx.doi.org/10.1006/jmre.2001.2474] [PMID: 11846583]

[20] L. Venkataramanan, Y.-Q. Song, and M.D. Hurlimann, "Solving Fredholm integrals of the first kind with tensor product structure in 2 and 2.5 dimensions", *IEEE Trans. Signal Process.,* vol. 50, no. 5, pp. 1017-1026, 2002. [http://dx.doi.org/10.1109/78.995059]

[21] I.K. Abu-Shumays, and L.D. Marinelli, "A smoothing solution (unfolding) of a two-dimensional density function from its spectrum", *J. Comput. Phys.,* vol. 7, p. 219, 1971. [http://dx.doi.org/10.1016/0021-9991(71)90086-6]

[22] P.T. Callaghan, and I. Furó, "Diffusion-diffusion correlation and exchange as a signature for local order and dynamics", *J. Chem. Phys.,* vol. 120, no. 8, pp. 4032-4038, 2004. [http://dx.doi.org/10.1063/1.1642604] [PMID: 15268569]

[23] P. Galvosas, Y. Qiao, M. Schönhoff, and P.T. Callaghan, "On the use of 2D correlation and exchange NMR spectroscopy in organic porous materials", *Magn. Reson. Imaging,* vol. 25, no. 4, pp. 497-500, 2007. [http://dx.doi.org/10.1016/j.mri.2006.11.009] [PMID: 17466772]

[24] P.T. Callaghan, S. Godefroy, and B.N. Ryland, "Use of the second dimension in PGSE NMR studies of porous media", *Magn. Reson. Imaging,* vol. 21, no. 3-4, pp. 243-248, 2003. [http://dx.doi.org/10.1016/S0730-725X(03)00131-0] [PMID: 12850714]

[25] V.V. Rodin, *Magnetic Resonance Methods.* Press MIPT: Moscow, 2004.

[26] R. Fechete, D.E. Demco, U. Eliav, B. Blümich, and G. Navon, "Self-diffusion anisotropy of water in sheep Achilles tendon", *NMR Biomed.,* vol. 18, no. 8, pp. 577-586, 2005. [http://dx.doi.org/10.1002/nbm.995] [PMID: 16320373]

[27] Y. Qiao, P. Galvosas, and P.T. Callaghan, "Diffusion correlation NMR spectroscopic study of anisotropic diffusion of water in plant tissues", *Biophys. J.,* vol. 89, no. 4, pp. 2899-2905, 2005. [http://dx.doi.org/10.1529/biophysj.105.064709] [PMID: 16100271]

[28] A. Traore, *Etude par RMM et diffusion des neutrons de la dynamique de l'eau dans le collagene nativ et dénaturé.* These. Universite Blaise Pascal, 1997.

[29] V.V. Rodin, and D.P. Knight, "Water self-diffusion in natural silk fibers as determined by the pulsed magnetic field gradient method", *Biophysics,* vol. 48, no. 3, pp. 404-410, 2003. [PMID: 12815853]

[30] M.A. Javed, P.M. Kekkonen, S. Ahola, and V-V. Telkki, "Magnetic resonance imaging study of water absorption in thermally modified pine wood", *Holzforschung,* vol. 69, no. 7, pp. 899-907, 2015. [http://dx.doi.org/10.1515/hf-2014-0183]

CHAPTER 7

PFG NMR in Studying Solutions of Carboxylated Acrylic Polymers

Abstract: The solutions of random copolymers composed from butyl methacrylate (BMA) and methacrylic acid (MAA) with variable molar ratio of BMA/MAA (between 100/0 and 60/40) in isopropanol (IPA) and IPA-water have been studied by the pulsed field gradient (PFG) nuclear magnetic resonance (NMR) method. The self-diffusion coefficients of isopropanol in BMA-MAA solutions were defined from the initial parts of echo-attenuation curves whereas the polymer diffusivities (two orders of magnitude lower than those for solvent) were separately measured applying solvent signal suppression (maximum gradient of 10 T/m). The aim was to study an influence of polymer concentration and BMA/MAA molar ratio on the diffusion constants of solvent and polymer. The diffusion coefficients for IPA increased with increasing BMA content in copolymer. The calculations based on Wang model to describe the solvent diffusion showed that relative mass of bound solvent per mass of polymer increased with increasing MAA content in the copolymer. The model of restricted polymer diffusion have been applied to explain the data on diffusivity of polymer. The macromolecules formed the transient aggregates with effective size that decreased with decreasing MAA content. Some polymer aggregates composed network and resulted in obstructive effect for diffusion of the rest nonconnected copolymer chains. At the study of BMA-MAA copolymers in IPA-water solutions, the focus was on the effects of molecular composition and neutralisation level of methacrylic acid.The results obtained showed that the average diffusion coefficient of solvent in D-IPA/D_2O solutions of neutralised copolymer was larger than that in un-neutralised copolymer IPA solutions. Polymer diffusion at infinite dilution in IPA-water allowed an estimate for the effective hydrodynamic radius of unperturbed diffusive polymer as 2.6-4.8 nm.

Keywords: Apparent diffusion coefficient (D_{app}), Bound solvent fraction (h_b), Butyl methacrylate (BMA), Deuterated isopropanol (D-IPA), Diffusion coefficient for solvent ($D_{solvent}$), Diffusion coefficient for polymer ($D_{polymer}$), Diffusion time Δ, Effective hydrodynamic radius of the polymers R_H, Isopropanol (IPA), Mass of bound solvent (m_b), Mass of polymer (m_p), Mean square displacement $<z^2>$, Methacrylic acid (MAA), Neutralised copolymer, Nuclear Magnetic Resonance (NMR), Pulsed Field Gradient (PFG), "Solvation" parameter ($H=m_b/m_p$), Triethanolamine (TEA), Weight fraction of polymer (w_p).

Victor V. Rodin

INTRODUCTION: CARBOXYLATED ACRYLIC POLYMERS

Polymers are now produced in great variety for different applications, in particular, for film packaging, adhesives, composites, long action drugs, paints and coatings [1 - 6]. The films and other formulations can be produced from solutions of polymers [1 - 3, 6 - 8]. Both experimental and theoretical studies of copolymer solution properties have been presented in literature [1, 5 - 11]. The selectivity of the solvent is often the main factor defining the structure in solution of block copolymers [1, 3 - 6, 12]. Many properties of random copolymers and industrial applications are dependent on the number of hydrophobic and hydrophilic monomers and the distribution of monomer sequences in the chain of macromolecules [1 - 3, 6 - 9, 11, 12]. Carboxylated acrylic polymers are important, for instance, in development of water-borne paints and delivery systems [3, 5 - 7, 12 - 14]. Polyelectrolyte amphiphilic copolymers can aggregate in many solvents and result in structured solutions. The sequence of monomers in macromolecular chain defines the characteristics of random copolymers, for example, composed from butyl methacrylate (BMA) and methacrylic acid (MAA) [2, 8, 10, 11]. Such copolymers could be made in the alcohol solutions, in particular, in isopropanol (IPA), *i.e.,* in solvents of intermediate polarity [2, 7, 10]. The glass transition temperature (T_g) of random copolymers is in the range from 27°C (mol% BMA = 100) to 128°C (mol% BMA = 60). Some NMR/MRI experiments on the role of monomer structure in the association process of polymers in solutions of BMA-MAA random copolymers in IPA have been carried out in Refs. [7, 8, 10 - 13]. Moreover, Ref. [8] presented the experimental data on random copolymers that comprise methylmethacrylate (MMA) instead MAA and *tert*-butyl methacrylate (*t*BMA) instead BMA, *i.e.*, copolymers MMA-*t*BMA at mol% MMA=50 (Fig. **1**). The T_g for this copolymer is equal to 100°C. Relative hydrophobicity of the copolymers MMA-BMA is comparable with that in MMA-*t*BMA [8].

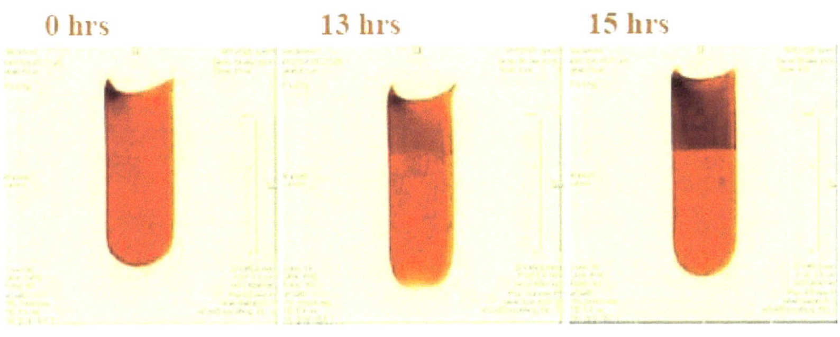

Fig. (1). Images of the MMA-*t*BMA copolymer (mol% MMA=50) in IPA: phase separation of copolymer solution appeared in few hours. ^{1}H MRI measurements (400 MHz, Bruker) have been performed with MSME pulse sequence at TE = 7.4 ms and T=297 K [8].

This chapter considers the PFG NMR studies of BMA-MAA copolymer solutions in IPA and in IPA-water analyzing the diffusion coefficients of solvent and macromolecules. The main purpose was to study the influence of polymer composition and neutralisation level of MAA on diffusion characteristics of these random copolymers in solutions.

SELF-DIFFUSION OF SOLVENT AND POLYMER IN SOLUTIONS OF RANDOM COPOLYMERS IN ISOPROPANOL

The BMA-MAA random copolymer solutions were prepared as described in [10]. The solvent was removed from the BMA-MAA copolymer solutions by drying. The dry polymer could be dissolved in the deuterated chloroform to record high-resolution ^1H NMR spectra and to identify the composition and structure [2, 7, 10]. Fig. (**2**) shows one example of such a spectrum recorded for pure poly(BMA). Additional NMR analysis have been done on the solutions of the BMA-MAA copolymers in deuterated IPA (DIPA) as it is shown in the Fig. (**3**). In order to prepare the BMA-MAA copolymers for studying the effect of neutralisation of MAA, the following procedure has been carried out. Firstly, the dry polymer was dissolved in DIPA at 50% solids. Triethanolamine (TEA) was added (1:1, 0.75:1 or 0.5:1 molar ratio of TEA to COOH in the polymer). TEA is a strong base. The samples with TEA were gently rolled for ~1 week. After that, water (D_2O) was added to the samples to get desired polymer concentration.

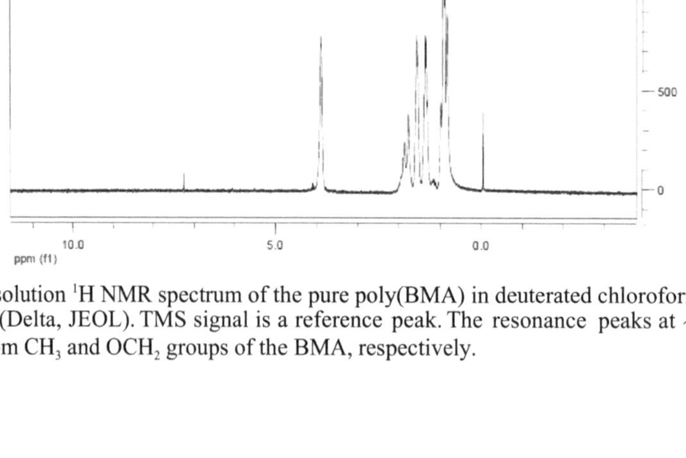

Fig. (2). High-resolution ^1H NMR spectrum of the pure poly(BMA) in deuterated chloroform ($CDCl_3$). T = 298 K. 270 MHz (Delta, JEOL). TMS signal is a reference peak. The resonance peaks at ~0.9-1.0 ppm and ~4.0 ppm are from CH_3 and OCH_2 groups of the BMA, respectively.

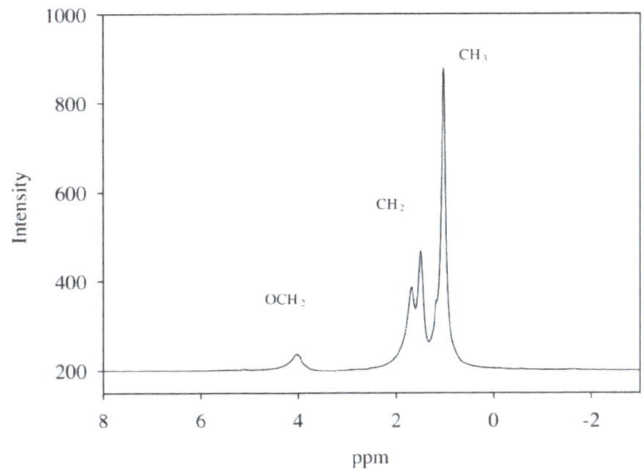

Fig. (3). ^{1}H NMR spectrum of the pure poly(BMA) (~32% solids) in D-IPA. T=298 K. 300 MHz (DSX 300, Bruker). Molecular weight of polymer is ~38 K.

^{1}H NMR spectroscopy has been applied to control molar content of BMA in the copolymers. Normally, the resonance signals of CH_3 groups (~1.0 ppm) and OCH_2 groups (~4.0 ppm) of BMA have been monitored with measuring their integral intensities. Fig. (**4**) presents one example of ^{1}H NMR spectrum of BMA-MAA random copolymers with neutralisation of MAA segments by TEA (1:1) and diluted with D_2O to get 5% solids in solution. The resonance peaks of polymer are shown on the spectrum. The proton signals at ~3.13 and ~3.8 ppm were responsible for the CH_2 groups in segments CH_2-N and CH_2-OH of TEA: $(CH_2OHCH_2)_3N$.

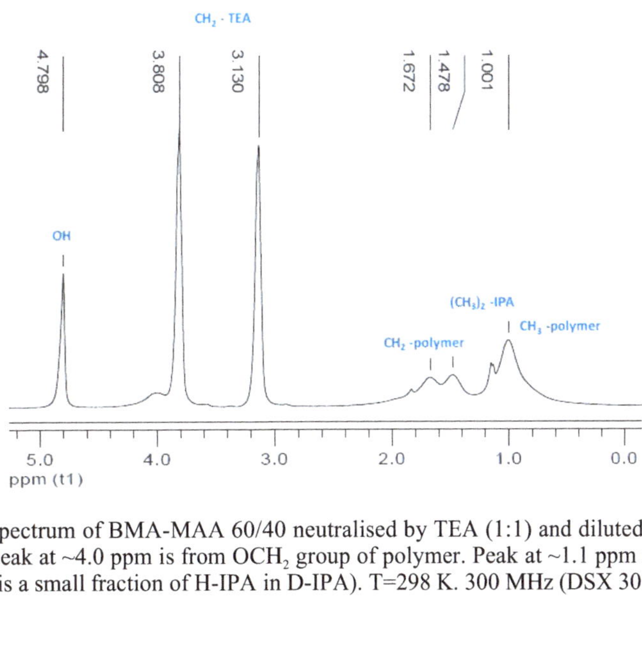

Fig. (4). ^{1}H NMR spectrum of BMA-MAA 60/40 neutralised by TEA (1:1) and diluted with D_2O to get 5% solids in solution. Peak at ~4.0 ppm is from OCH_2 group of polymer. Peak at ~1.1 ppm is from $(CH_3)_3$ group of solvent (as there is a small fraction of H-IPA in D-IPA). T=298 K. 300 MHz (DSX 300, Bruker).

NMR diffusion experiments were carried out on a DSX 300 NMR instrument (Bruker) using a diff30 probe. Stimulated-echo sequence of the three $\pi/2$ pulses was applied [2, 11, 15 - 18] (Chapter **1**, Fig. **7**, top). In this pulsed sequence, the time interval τ_1 presents the gap between the 1st and 2nd $\pi/2$ pulses whereas the interval τ_2 is the gap between the second and third ones. The pulsed sequence resulted in a formation of stimulated echo at time $2\tau_1+\tau_2$. At studying solvent diffusion, the intensity of magnetic field gradient (amplitude *G*) was increased linearly up to maximal value 1 T/m. When the low values of diffusion constant for polymers have been studied, solvent suppression was used. The maximum value of *G* was 10 T/m. A nonlinear least squares fitting procedure has been applied to the experimental data to calculate the diffusion constants. OH peak intensities have been used in calculations of solvent diffusion constants $D_{solvent}$ [2, 10, 11]. The intensities of the CH_3 peak (~1 ppm) in the spectra measured at different gradient values in diffusion experiment have been used to obtain the diffusion constants for polymer $D_{polymer}$ [10, 11]. The value τ_1 in all the diffusion experiments was 2.54 ms. The Stejskal-Tanner relationship [15 - 19] between the echo amplitude, gradient factor $b=(\gamma G\delta)^2\cdot(\Delta-\delta/3)$, and diffusion constant *D* has been used according to Chapter **1**, eq. (6).

The works [10, 11] presented the NMR diffusion experiments on the BMA-MAA random copolymers in IPA in the range of solids 0–32%. Those data showed the approaches how to measure the diffusion constants of solvent molecules and polymers separately (solvent diffusivity was much higher than that for polymer) in the same diffusion experiment. The IPA diffusion coefficients increased with molar content of BMA in the copolymers. Different models describing molecular diffusion in polymer systems have been considered [2, 8 - 12, 20 - 36]. Using the approach with Wang's model [10, 11, 25], it was possible to determine a bound solvent component which increased with increasing the molar content of MAA. It was found that the diffusion constants for solvent in IPA solutions of the BMA-MAA random copolymers are significantly less than that for pure IPA. This suggested [10] that solvent experiences a slower diffusion in the copolymer solution than in IPA. The results revealed [10 - 13] that the diffusion constant of polymer increased with BMA content. The data showed also that an effective hydrodynamic radius decreased with increasing BMA content. The data on polymer diffusion at infinite dilution in IPA resulted in an estimate of the effective hydrodynamic radius in the range of 5–6 nm [11]. The diffusion constants of macromolecules increased with BMA content. Thus, the effective size of the polymer aggregates decreased with increasing BMA segments. The model of anomalous restricted diffusion of macromolecules has been accepted to explain the NMR diffusion data on BMA-MAA copolymers in IPA [11]. Two types of polymer chains were responsible for the anomalous diffusion behavior. One type of chains was considered as the transient aggregates. Other chains could move through these aggregates. Thus, the aggregates form a network and create obstructing effects for movement of the chains of second type. An exchange between these two states of the polymer within the diffusion time resulted in observable diffusion characteristics. A variation of the MAA (BMA) content influences the extent of diffusion restriction in the solutions of the BMA-MAA copolymers. This affects the motions of polymer aggregates as well as an exchange between mobile polymer chains and immobile polymer network in IPA. A replacement of isopropanol by IPA-H_2O mixtures in the BMA-MAA solutions can lead to a more rich behavior of macromolecules and interactions in polymer solution.

EFFECT OF POLYMER CONCENTRATION AND BMA/MAA MOLAR RATIO ON HYDROGEN BONDING IN NEUTRALISED COPOLYMERS

Fig. (**4**) shows the ^{1}H NMR spectrum of BMA-MAA D-IPA/TEA solution diluted with D_2O. These data show that there is one OH peak because of exchange between OH/OD from IPA and OH from MAA segments. When D-IPA was used, then only small residual OH signal could be available in IPA. A chemical shift of the OH signal in the spectrum was dependable on level of dilution and mol% MAA in copolymer. Fig. (**5**) shows few examples on chemical shift of the OH signal in neutralised copolymers *vs.* MAA content and polymer concentration.

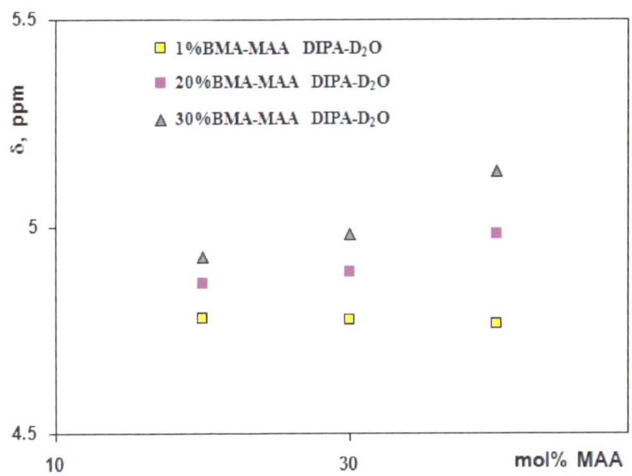

Fig. (5). Chemical shift of OH peak in neutralised BMA-MAA 60/40 copolymer (TEA: COOH=1:1) in D-IPA/D_2O (polymer concentrations are 1, 20 and 30%) as a function of mol% MAA. T=298 K. 300 MHz (DSX 300, Bruker).

The published data on un-neutralised BMA-MAA copolymers in IPA showed [7, 10] that a value of chemical shift for OH resonance was depending on the strength of H-bonds. Those data presented the ^{1}H NMR spectra from the poly(BMA)-IPA solutions diluted with $CDCl_3$. When the IPA content was decreased, the OH peak in the spectrum of poly(BMA) in IPA shifted to higher field. The data showed [10] that on the dilution with $CDCl_3$, *i.e*., when the IPA content decreased, the resonance of the OH group (originally at ~4.7 ppm) shifted to higher field. This confirmed that in both pure IPA and in the poly(BMA)-IPA solutions H-bonds are destroyed at increasing $CDCl_3$ content [10].

A magnetic shield for proton of OH group is different in associated (H-bonded) and un-associated molecules. OH signal will be registered in the field with chemical shift which is mean with necessary weights for two edge states. Some factors, in particular, temperature and dilution change a population of associated and un-associated states. Thus, OH chemical shift is a function of temperature and concentration. A dilution of the BMA-MAA IPA solutions by solvent without ability to form H-bonds result in dissociation of alcohol molecules [10, 11].

MAA monomers of BMA-MAA copolymer contain COOH groups that participate in additional H-bonding. Therefore, much stronger H-bonding is observable at higher polymer content. Moreover, H-bonding is stronger at larger mol% MAA. That is why OH resonance in the copolymer with mol% MAA=60 is stronger shifted to weak field (large value of δ is registered) than that in the copolymers with low mol% MAA. The process of neutralisation of MAA groups decreases the effect of OH shifting to weak field. For the BMA-MAA copolymers in IPA (*i.e*., for un-neutralised copolymers, Fig. **6**) the shift of OH resonance to the region of large chemical shifts was quite strong.

When polymer concentration increases, the IPA content decreases and OH peak shifts to the right field of the spectrum (to smaller δ values; H-bonds are partly broken and not much associated isopropanol molecules). This tendency is distinctly seen in the pure poly(BMA) in IPA solution (Fig. **6**). Second tendency is a shift of OH peak (MAA segments of copolymer participates in H-bonding) to weak field (to left area of the spectrum, where a value of chemical shift becomes larger) at increase of polymer concentration. Thus, a participation of additional H-bonds in forming the large-scale associates has a place. The resulting OH chemical shift characterises a combination of two mentioned above states.

Fig. (**7**) compares the chemical shifts of neutralised BMA-MAA copolymers (TEA:COOH=1:1) diluted with D_2O and un-neutralised BMA-MAA copolymers in IPA. This comparison shows that at the same polymer content (*e.g.,* 30% solids) OH resonances in un-neutralised copolymers are more shifted to weak field than those in neutralised copolymers (Figs. **5**, **6**, **7**). The chemical shift values have the big changes with mol% MAA. The data show as well that for diluted solutions (when solids ~1–5%), a value of OH chemical shift is practically constant (non depending on the mol% MAA). In such a case, the chemical shift of OH peak is rather closed to that in water (Fig. **6**).

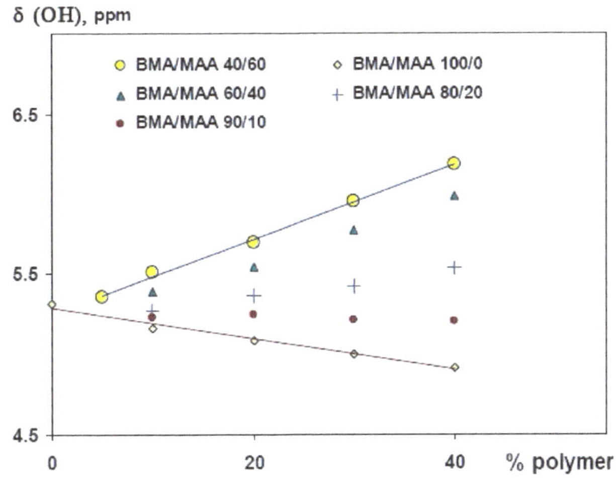

Fig. (6). Chemical shift of OH peak in copolymer solutions (BMA-MAA in D-IPA) as a function of polymer concentration. For the BMA-MAA 100/0 OH signal is responsible for residual OH groups of IPA in deuterated IPA. T=298 K. 300 MHz (DSX 300, Bruker).

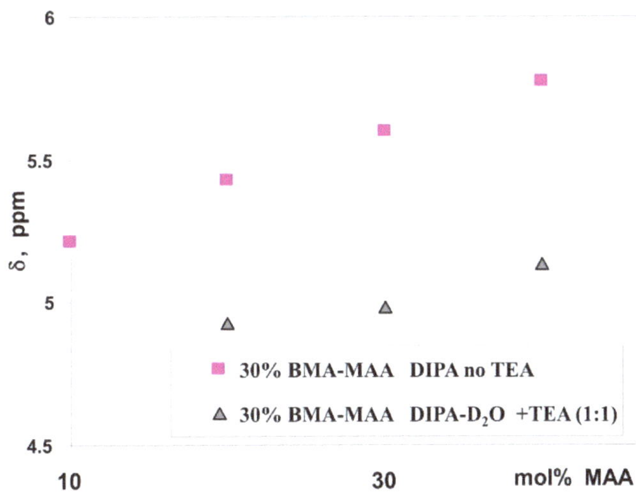

Fig. (7). Chemical shifts of OH peak in un-neutralised BMA-MAA (diluted with IPA) and neutralised (TEA:COOH=1:1) BMA-MAA (diluted with D_2O) copolymers at 30% solids *vs.* mol% MAA. T=298 K. 300 MHz (DSX 300, Bruker).

Fig. (**8**) shows chemical shifts of OH peaks in neutralised (diluted with water) and un-neutralised (diluted with IPA) 80/20 and 60/40 BMA-MAA copolymer solutions as dependences on polymer concentration. These data discovered distinct sharp rise of δ when polymer content exceeds 25-30%. At this polymer concentration, there is already significant association of neutralised BMA-MAA copolymer chains [8, 11 - 13]. This is in line with viscosity data [7, 37 - 39]. Maximum of viscosity for the neutralised BMA-MAA copolymers has been obtained in the range of polymer concentration 25-30%.

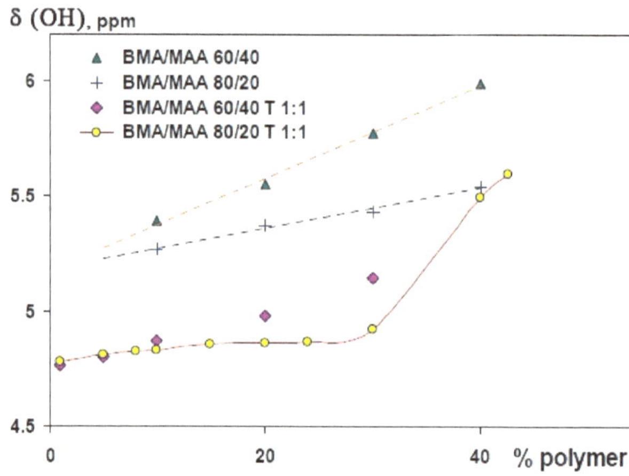

Fig. (8). Chemical shift δ of OH peak in neutralised (TEA:COOH=1:1) and un-neutralised copolymer solutions (BMA/MAA 80/20 and 60/40) as a function of polymer concentration. T=298 K. 300 MHz (DSX 300, Bruker).

Fig. (**9**) shows the chemical shifts of OH resonance in neutralised copolymer solutions (BMA-MAA 70/30 copolymers in D-IPA/D_2O) as a function of polymer concentration at variable neutralisation level (TEA:COOH = 1:1; 0.75:1 and 0.5:1).

The data obtained on the copolymers with mol% BMA = 60, 70 and 80 show a very weak dependence of OH chemical shift on the polymer concentration in the range of < 25%. However, at higher concentration of polymer (~30% and higher), a value of chemical shift δ(OH) sharply increased showing significant role of COOH groups (MAA-monomers) in additional H-bonding in this concentration range. This effect looks stronger for the copolymers with the smallest mol% BMA = 60 from the studied copolymers, *i.e.*, at largest MAA content. The data for the samples with various level of neutralisation of the copolymers (at the same polymer concentration) show that δ of OH peak is larger for higher neutralisation level (TEA:COOH=1:1, *e.g.*, Fig. **9**). Thus, in the range of polymer concentration > 25–30%, strong H-bonding resulting in big δ values is obtained for fully neutralised copolymers.

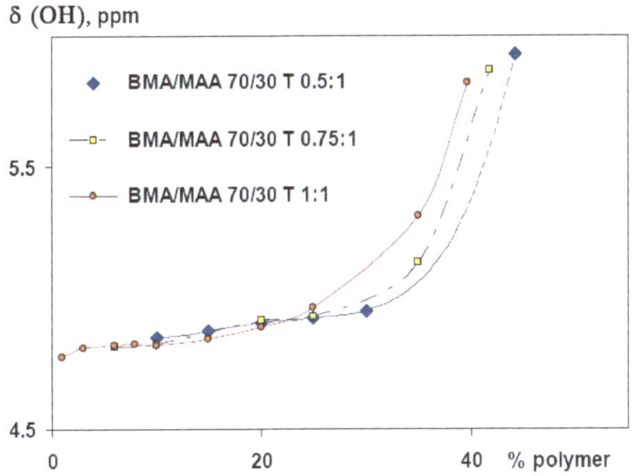

Fig. (9). Chemical shift of OH peak in solutions of neutralised copolymers (BMA-MAA 70/30 copolymers in D-IPA/D_2O) as a function of % polymer at variable neutralisation level (TEA:COOH=1:1; 0.75:1 and 0.5:1). T=298 K. 300 MHz (DSX 300, Bruker).

SOLVENT AND POLYMER DIFFUSION STUDIES IN NEUTRALISED COPOLYMERS

Solvent Diffusion: Effect of mol% BMA in BMA-MAA Random Copolymers and Polymer Concentration at Neutralisation Level TEA:COOH = 1:1

A typical echo-attenuation plot in NMR diffusion experiment for OH peak at diffusion time Δ=50 ms could be analysed as a bi-exponential fit for two distinct fractions of protons [10]. The isopropanol diffusion constant is calculated from initial (fast decaying) slope of this attenuation plot. The slowly decaying part of the curve is assigned to the protons of the macromolecules because of low polymer diffusivity [11]. For example, for the BMA-MAA 80/20 copolymers (diluted in IPA) in the range of 5-15% polymer a difference between solvent and polymer diffusion constants is a factor of 30–80). D_{app} for the polymer have been measured during solvent suppressed experiment when the gradient factor $b = (\gamma G\delta)^2\times(\Delta-\delta/3)$ exceeded 0.3×10^{11} $rad^2 s m^{-2}$ (*i.e.*, when gradient G is in the range 2.5 T/m $\leq G \leq G_{max}$ = 10 T/m).

Fig. (**10**) compares the solvent diffusion constants for the neutralised BMA-MAA copolymers in D-IPA/D_2O presented as functions of % solids at mol% BMA = 60, 70 and 80. The diffusion coefficients increase with increasing BMA content (at constant concentration of the macromolecules). Similar tendency was discovered for diffusion coefficients of BMA-MAA random copolymers in IPA in the publication [10]. The data obtained show that OH groups of IPA interact with the MAA segments and the OH-bonding with the solvent leads to a more expanded polymer conformation with increasing the MAA content. A comparison of the data for the BMA-MAA copolymer in D-IPA/D_2O with those for copolymers in IPA [10, 11] shows that in fast exchange regime between OH groups of solvent and MAA units the average solvent diffusion coefficient is larger for neutralised copolymer in D-IPA/D_2O than for un-neutralised (in IPA) copolymers. One particular example (solids 15%) shows that $D_{solvent}$=9.98×10^{-10} m^2/s in neutralised BMA-MAA 80/20 random copolymer solution (in D-IPA/D_2O), whereas in un-neutralised BMA-MAA 80/20 copolymer solution (in IPA), $D_{solvent}$=4.05×10^{-10} m^2/s, and this is consistent with much stronger H-bonding (OH chemical shift data) in un-neutralised copolymers in comparison with neutralised BMA-MAA (Fig. 7).

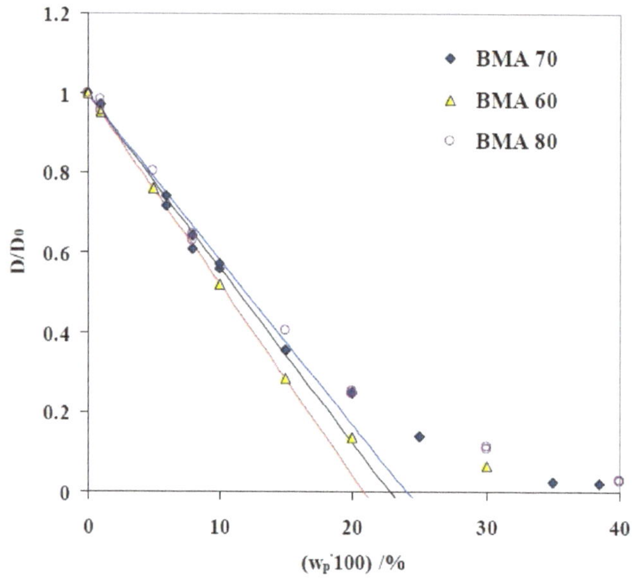

Fig. (10). Relative solvent self-diffusion constants as the functions of polymer weight fraction w_p for neutralised BMA-MAA random copolymers in D-IPA/D_2O at different molar BMA content (mol% =60, 70 and 80). D_0 is self-diffusion constant of solvent at infinite dilution of polymer in solution. The lines are the fits according to the Wang's equation [25] (presented also as eq. (4) in [10]) in range of 0–15% solids. T=298K. 300 MHz (DSX 300, Bruker).

In order to evaluate the effect of copolymer concentration and mol% BMA on solvent diffusion coefficients in neutralised copolymers and compare those for un-neutralised copolymers, the range between 0 and 15% polymer could be analysed. In this concentration range, self-diffusion coefficient of solvent decreased as a straight line with increasing the polymer concentration, and viscosity in neutralised samples was comparable with that in un-neutralised ones [37, 38]. To fit the data, the interaction between the solvent and the polymer should be taken into account. The model given by Wang [10, 25] can be applied again as it was used before for un-neutralised copolymers in IPA (eq.(4) in Ref. [10]).

In Wang's model, solvent molecules diffuse around the obstructing macromolecules resulting in the obstruction effect. It is considered also a "solvation effect" which defines the interaction between solvent and polymer. An exchange between bound and free states results in an averaged diffusion coefficient D which depends on the concentration of the polymer solution. In particular, in the dilute regime the randomly distributed macromolecules influence the solvent motion independently. Two factors are considered for the detail description [25, 26]. First parameter is a shape factor, α. This factor arises from the obstruction effect, and α=1.5 [10, 25 - 27]. The next one is a "solvation" parameter, H. It is describing the fraction of solvent immobilized at any instant by association with the macromolecule. In Wang's model, the composition is expressed by the polymer weight fraction w_p which is the dependent experimental variable. D_0 is the diffusion coefficient of pure solvent. V_{sp} is the specific volume of polymer, and V_{ss} is the specific volume of solvent. The "solvation" parameter $H = m_b/m_p$ is the mass of bound solvent m_b per mass of polymer m_p.

The plot of D/D_0 *versus* polymer weight fraction w_p (Fig. **10**) gives a linear dependence (at 0–15% polymer), and the solid line is calculated for Wang's model [10, 25]. A linear fit gives an intercept with the concentration axis to define the w_p values. For neutralised BMA-MAA 70/30 copolymer, a linear fit gives an intercept with the concentration axis at w_p = 0.228 (polymer concentration 22.8%). The value w_p in neutralised BMA-MAA copolymers changed from 0.241 to 0.209 when the mol% MAA in the BMA-MAA copolymer increased from 20 to 40%. This w_p range (*i.e.* polymer concentrations which result in $D/D_0 = 0$) is much larger for un-neutralised BMA-MAA 60/40 and 80/20 copolymers [10]. In the case of un-neutralised BMA-MAA copolymers, the value w_p changed from 0.58 to 0.44, when the mol% MAA in the BMA-MAA copolymer increased from 20 to 40%. The interaction between the OH groups of the solvent and the MAA segments is stronger than the hydrophobic interaction between the CH_3 groups of both solvent and macromolecules. This explains why the solvent diffusivity in the BMA-MAA 60/40 copolymer solution resulted in the intercept with the concentration axis at lower polymer weight fraction w_p than that in the BMA-MAA 80/20 solution.

In the range of 20–25% polymer a solvent diffusivity (as a function of polymer concentration) changed its slope. As it was mentioned above, this polymer concentration (25% solids) of neutralised BMA-MAA copolymers in IPA-water already results in sharp increase of viscosity [12, 38]. Wang's model (considering obstruction and solvation effects) probably is not suitable to fit the data for neutralised copolymer in this range of polymer concentration.

Polymer Diffusion: Effect of BMA/MAA Molar Ratio at Neutralisation Level TEA:COOH=1:1

Fig. (**11**) presents the polymer diffusion data for the range of copolymer samples. Although data obtained are scattered for some polymer concentrations (less than 20%) there is a trend of increasing diffusion coefficient with increasing BMA content. This trend has been observed above also in solvent diffusion data. Chemical shift data showed as well that extent of H-bonding with solvent increased at the increase of MAA content.

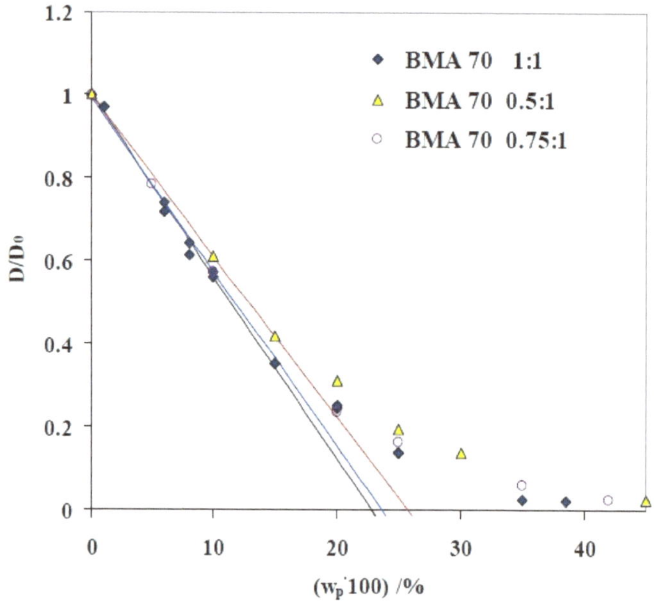

Fig. (11). Polymer diffusion constants D_{app} as the functions of polymer weight fraction w_p for neutralised BMA-MAA random copolymers in D-IPA/D_2O at different mol% BMA. TEA:COOH = 1:1. The solid lines through sets of data for mol% BMA = 80 (diamonds) and for mol% BMA = 70 (triangles) represent the fits using the formula $D{=}D_{0p}{\times}(1-\phi{\times}w_p)$. These estimate roughly diffusion constants D_{0p} at infinite dilution. ϕ is a friction coefficient. T=298 K. 300 MHz (DSX 300, Bruker).

According to Fig. (**11**), polymer self-diffusion coefficient decreased with increasing the polymer concentration. If the polymer molecules are considered as spherical particles, then the decrease in *D* values can be described using the Stokes–Einstein equation [11]. In this case, the effective hydrodynamic radius of the polymers is seemed to increase with increasing the polymer concentration or mol% MAA. This testifies about the solvation of the hydrophilic parts of the polymer.

As the diffusion coefficient decreases with increasing concentration in the range 0–20%, a simple formula $D = D_{op}{\times}(1 - \phi\ {\times}w_p)$ has been tried to fit the data

(Fig. **11**). This linear fit gives an estimate D_{op} of polymer at infinite dilution. Using the Stokes-Einstein equation, one can estimate the effective hydrodynamic radius R_H of the polymers [11]. These roughly R_H values are: ~ 4.79 nm (mol% BMA = 70), ~ 2.69 nm (mol% BMA = 80). Thus, an average size R_H of polymer particles estimated from the diffusivity of the unperturbed polymer is ~ 3.7 ± 1.0 nm. In the case of un-neutralised BMA-MAA copolymers (diluted in IPA), an average size R_H of the unperturbed polymer in IPA is ~ 6.1 ± 0.5 nm [11].

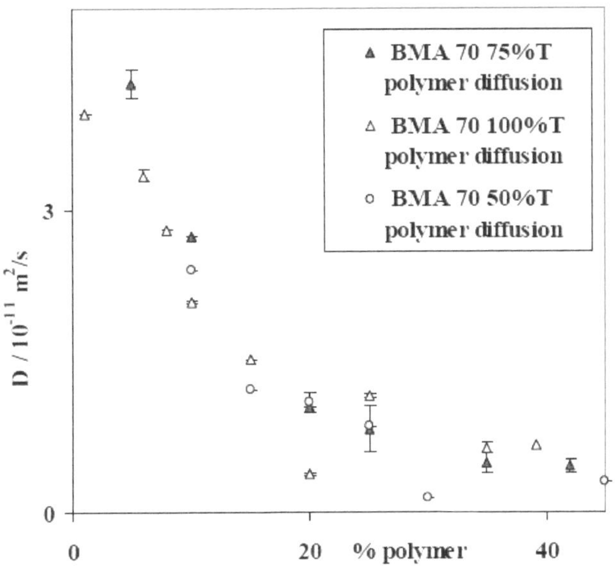

Fig. (12). The relative self-diffusion constants D/D_0 of the solvent vs polymer weight fraction w_p for BMA-MAA 70/30 random copolymer in D-IPA/D_2Oat various neutralisation level (TEA:COOH = 1:1, 0.75:1, and 0.5:1). D_0 is self-diffusion constant of pure IPA. Polymer concentration = w_p.100%. The lines are the fits to the Wang's equation [25] for the data in the range of 0-15% solids. T=298 K. 300 MHz (DSX 300, Bruker).

Polymer and Solvent Diffusion: Different Neutralisation Level (TEA:COOH = 1:1; 0.75:1 and 0.5:1)

Fig. (**12**) compares the solvent diffusion coefficients in neutralised BMA-MAA 70/30 random copolymer solutions (diluted in D-IPA/D_2O) for various neutralisation level. The diffusion coefficients slightly increase with decreasing neutralisation level from 100% to 50% (at constant concentration of the polymer). The diffusion constants decreased with increasing the polymer concentration. In the range of polymer concentration between 0 and 15% solids, the analysis of solvent diffusion discovered the linear dependences of the diffusivities on polymer content (Fig. **12**). Therefore, Wang's equation [25] was applicable to determine an intercept of the lines with the concentration axis to give the w_p

values and estimate solvation parameter H [10].

For fully (100%) neutralised BMA-MAA 70/30 copolymer a linear fit gives an intercept with the concentration axis at $w_p = 0.228$ (solids 22.8%). The value w_p in partly neutralised BMA-MAA 70/30 random copolymers (Fig. **12**) changed from 0.258 (TEA:COOH=0.5:1) to 0.237 (TEA:COOH = 0.75:1). For BMA-MAA 80/20 neutralised copolymers similar treatment and linear fits gave the value w_p at $D/D_0 = 0$ as follows: 0.241 (at TEA:COOH = 1:1) and 0.275 (at TEA:COOH = 0.5:1). Thus, this is showing the same tendency for both BMA-MAA (mol% BMA=70) and BMA-MAA (mol% BMA=80) copolymers: an increase of w_p for non-fully neutralised polymer. This is consistent with un-neutralised copolymer data, when w_p value was quite big, *e.g.*, 0.58 in BMA-MAA 80/20 copolymer in IPA [10].

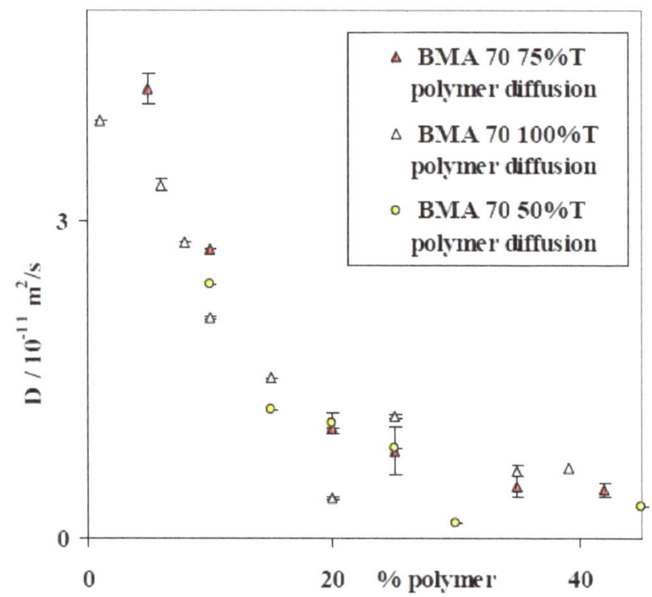

Fig. (13). Polymer diffusion constants D_{app} as a function of polymer concentration for neutralised BMA-MAA 70/30 random copolymers in D-IPA/D_2O at different neutralisation level (TEA:COOH=1:1, 0.75:1 and 0.5:1). T=298 K. 300 MHz (DSX 300, Bruker).

Figs. (**13** and **14**) show the polymer diffusion data for BMA-MAA (70/30 and 80/20) copolymer samples at different neutralisation level (TEA:COOH=1:1; 0.75:1 and 0.5:1). A diffusion coefficient decreased with increasing the polymer concentration. Considering polymer molecules as spherical particles, it is possible to describe the decrease in D values using the Stokes–Einstein equation [11, 35]. In this case, the effective hydrodynamic radius of the polymers is seemed to increase with increasing the polymer concentration.

As it was estimated above for the fully neutralised BMA-MAA copolymers (mol% BMA=80, 70, 60) in IPA-water (Fig. **11**), the effective hydrodynamic radius of the polymers is seemed to increase with increasing mol% MAA, *e.g.*, from 2.69 nm (mol% MAA=20) to 4.79 nm (mol% MAA=30). This is presumably a result of the solvation and expansion of the hydrophilic parts of the chain. The linear fits in the range of 0–15% solids for the data of the BMA-MAA 80/20 and 70/30 copolymers in IPA-water with different level of neutralisation (Figs. **13** and **14**) showed similar results: the effective hydrodynamic radius R_H of the variously neutralised copolymers at infinite dilution is roughly estimated in the same range ~ 2.5–5 nm.

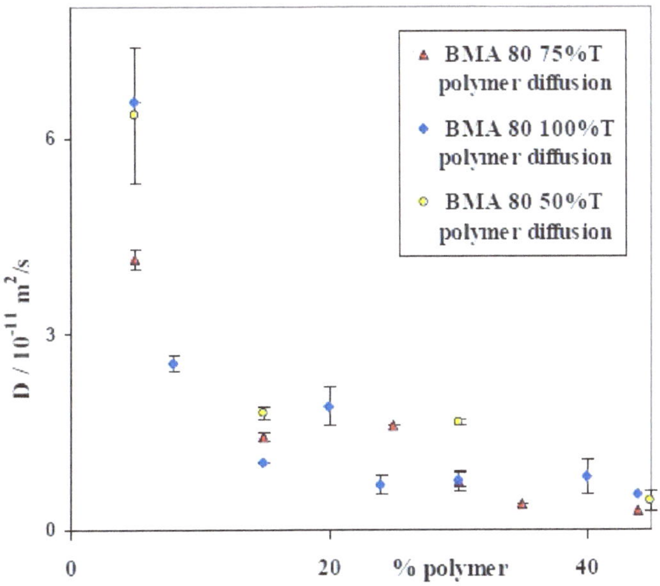

Fig. (14). Diffusion coefficients for polymer in BMA/MAA 80/20 random copolymer solutions *vs* polymer concentration at different neutralisation level (TEA:COOH=1:1; 0.75:1 and 0.5:1). T=298 K. 300 MHz (DSX 300, Bruker).

CONCLUDING REMARKS

The chapter considered ^{1}H NMR chemical shifts studies and PFG NMR diffusion experiments on the BMA-MAA copolymer solutions in IPA and IPA-water at various concentrations (up to 40% solids) and variable mol% BMA. The neutralised copolymers are strongly associated at polymer concentration ≥ 25%. OH chemical shift sharply increased in this concentration range. The interactions between the MAA units and the OH groups of solvent are stronger than the hydrophobic association between the CH_3 groups of macromolecules and solvent. H-bonding is much stronger in un-neutralised copolymers than in neutralised ones.

NMR data for diffusion of solvent and polymer could be separately obtained in one experiment. A maximum value of G_{max} = 10 T/m for the gradient pulse was used to suppress solvent signal and to measure the low value of the diffusion coefficient for the polymer. The diffusivity for polymer is much slower than that for solvent - by a factor of 30–80. The diffusion coefficients for solvent increased with mol% BMA.

Diffusion coefficients determined for different molar ratios BMA/MAA in the copolymers and for various polymer contents revealed the existence of exchangeable interaction with polymer units. The diffusion constants of solvent in neutralised copolymer solutions in D-IPA/D_2O were larger than those in un-neutralised copolymer IPA solutions.

The results obtained showed that diffusion constants of polymer increased with mol% BMA, and an effective hydrodynamic radius decreased with increasing mol% BMA. Polymer diffusion at infinite dilution of BMA-MAA copolymer solutions in IPA-water gave an estimate for the effective hydrodynamic radius R_H of unperturbed diffusive polymer in the range of 2.6–4.8 nm. This size was not roughly sensitive to different neutralisation level.

ACKNOWLEDGEMENTS

The author thanks Terence Cosgrove and Martin Murrey.

REFERENCES

[1] T.H. Epps, and R.K. O'Reilly, "Block copolymers: controlling nanostructure to generate functional materials – synthesis, characterization, and engineering", *Chem. Sci. R. Soc. Chem,* vol. 7, no. 3, pp. 1674-1689, 2016.

[2] V.V. Rodin, and T. Cosgrove, "Nuclear magnetic resonance study of water-polymer interactions and self-diffusion of water in polymer films", *OALJ: Chem. Mater. Sci,* vol. 3, no. 10, pp. 1-17, 2016.

[3] P. Ravi, C. Wang, K.C. Tam, and L.H. Gan, "Association behavior of poly(methacrylic acid)-*block*-poly(methyl methacrylate) in aqueous medium: potentiometric and laser light scattering studies", *Macromolecules,* vol. 36, no. 1, pp. 173-179, 2003. [http://dx.doi.org/10.1021/ma021302j]

[4] Y. Liu, A. Gajewicz, V.V. Rodin, W-J. Soer, J. Scheerder, G. Satgurunathan, P.J. McDonald, and J.L. Keddie, "Explanations for water whitening in secondary dispersion and emulsion polymer films", *J.Polym. Sci., B, Polym. Phys.,* vol. 54, no. 16, pp. 1658-1674, 2016. [http://dx.doi.org/10.1002/polb.24070]

[5] M. Mizusaki, Y. Shimada, Y. Morishima, and S. Yusa, "pH-Responsive intra- and inter-molecularly micelle formation of anionic diblock copolymer in water", *Polymers (Basel),* vol. 8, no. 2, p. 56, 2016. [http://dx.doi.org/10.3390/polym8020056]

[6] M. Monroy-Barreto, M.D.C. Perez-Vazquez, M. Briseno-Teran, and N. Esturau-Escofet, "Micro-structural characterization of diblock copolymers formed by styrene and different methacrylic units", *Int. J. Polym. Analys. Characterization,* vol. 19, no. 1, pp. 22-30, 2014.

[7] V.V. Rodin, T. Cosgrove, M. Murray, and R. Buscall, "Intermolecular interactions upon film form-

ation from copolymer solutions", *Book of Abstracts XVIIth European Chemistry at Interfaces Conference,* 2005, p. 107, Loughborough.

[8] V.V. Rodin, T. Cosgrove, and M. Murrey, "Magnetic resonance imaging and pulsed field gradient NMR in study of structure of polymer solutions and films", *Proceedings of the 6th Colloquium on Mobile Magnetic Resonance,* 2006, pp. 32-33, Aachen, Germany.

[9] V. Bütün, S.P. Armes, N.C. Billingham, Z. Tuzar, A. Rankin, J. Eastoe, and R.K. Heenan, "The remarkable "flip-flop" self-assembly of diblock copolymer in aqueous solution", *Macromolecules,* vol. 34, no. 5, pp. 1503-1511, 2001. [http://dx.doi.org/10.1021/ma0018392]

[10] T. Cosgrove, V.V. Rodin, M. Murray, and R. Buscall, "Self-diffusion in solutions of carboxylated acrylic polymers as studied by pulsed field gradient NMR. 1. Solvent diffusion studies", *J. Polym. Res.,* vol. 14, no. 3, pp. 167-174, 2007. [http://dx.doi.org/10.1007/s10965-006-9087-1]

[11] T. Cosgrove, V.V. Rodin, M. Murray, and R. Buscall, "Self-diffusion in solutions of carboxylated acrylic polymers as studied by pulsed field gradient NMR. 2. Diffusion of macromolecules", *J. Polym. Res.,* vol. 14, no. 3, pp. 175-180, 2007. [http://dx.doi.org/10.1007/s10965-006-9088-0]

[12] V.V. Rodin, T. Cosgrove, C. Flood, M. Murray, and R. Schweins, "Hydrophobic/hydrophilic random copolymers in solutions", *ILL Experimental Report. N 9-11-1186. France, .*

[13] V.V. Rodin, T. Cosgrove, U. Keiderling, J. Rowe, and M. Murray, "Random copolymers based on MMA", *BENSC Experimental Report CHE-04-1377. Germany,* pp. 116-117, 2007.

[14] P. Broz, "Polymer-Based Nanostructures: Medical Applications", In: *Royal Society of Chemistry: London*, 2010. [http://dx.doi.org/10.1039/9781847559968]

[15] J.E. Tanner, and E. Stejskal, "Restricted self-diffusion of protons in colloidal systems by the pulsed-gradient, spin-echo method", *J. Chem. Phys.,* vol. 49, no. 4, pp. 1768-1777, 1968. [http://dx.doi.org/10.1063/1.1670306]

[16] J.E. Tanner, "Transient diffusion in system partitioned by permeable barriers. Application to NMR measurements with a pulsed field gradient", *J. Chem. Phys.,* vol. 69, no. 4, pp. 1748-1754, 1978. [http://dx.doi.org/10.1063/1.436751]

[17] P.T. Callaghan, *Principles of Nuclear Magnetic Resonance Microscopy.* Clarendon Press: Oxford, 2001.

[18] V.V. Rodin, "Methods of Magnetic Resonance in Studying Natural Biomaterials", In: *Encyclopedia of Physical Organic Chemistry,* Zerong Wang, Ed., Chapter 53. vol. 4, part 4. John Wiley & Sons, Inc., 2017, pp. 2861-2908. http://eu.wiley.com/WileyCDA/WileyTitle/productCd-1118470451.html

[19] V.V. Rodin, *Magnetic Resonance Methods.* Press MIPT: Moscow, 2004. ISBN:789 5-7417-0228-7.

[20] J.S. Mackie, and P. Meares, "The diffusion of electrolytes in a cation-exchange resin membrane. 1.Theoretical", *Proc. R. Soc. Lon. Ser-A,* vol. 232, 1955no. 1191, pp. 498-509

[21] L. Masaro, and X.X. Zhu, "Physical models of diffusion for polymer solutions, gels and solids", *Prog. Polym. Sci.,* vol. 24, pp. 731-775, 1999. [http://dx.doi.org/10.1016/S0079-6700(99)00016-7]

[22] R. Knauss, J. Schiller, G. Fleischer, J. Kärger, and K. Arnold, "Self-diffusion of water in cartilage and cartilage components as studied by pulsed field gradient NMR", *Magn. Reson. Med.,* vol. 41, no. 2, pp. 285-292, 1999. [http://dx.doi.org/10.1002/(SICI)1522-2594(199902)41:2<285::AID-MRM11>3.0.CO;2-3] [PMID: 10080275]

[23] X. Gao, and W. Gu, "A new constitutive model for hydration-dependent mechanical properties in

biological soft tissues and hydrogels", *J. Biomech.,* vol. 47, no. 12, pp. 3196-3200, 2014. [http://dx.doi.org/10.1016/j.jbiomech.2014.06.012] [PMID: 25001202]

[24] R. Bai, P.J. Basser, R.M. Briber, and F. Horkay, "NMR water self-diffusion and relaxation studies on sodium polyacrylate solutions and gels in physiologic ionic solutions", *J. Appl. Polym. Sci.,* vol. 131, no. 6, p. 40001, 2014. [http://dx.doi.org/10.1002/app.40001] [PMID: 24409001]

[25] J.H. Wang, "Theory of the self-diffusion of water in protein solutions. A new method for studying the hydration and shape of protein molecules", *J. Am. Chem. Soc.,* vol. 76, no. 19, pp. 4755-4763, 1954. [http://dx.doi.org/10.1021/ja01648a001]

[26] P.T. Callaghan, and J. Lelievre, "The influence of polymer size and shape on self-diffusion of polysaccharide and solvent", *Anal. Chim. Acta,* vol. 189, pp. 145-166, 1986. [http://dx.doi.org/10.1016/S0003-2670(00)83720-4]

[27] A. Gottwald, L.K. Creamer, P.L. Hubbard, and P.T. Callaghan, "Diffusion, relaxation, and chemical exchange in casein gels: a nuclear magnetic resonance study", *J. Chem. Phys.,* vol. 122, no. 3, p. 34506, 2005. [http://dx.doi.org/10.1063/1.1825383] [PMID: 15740208]

[28] M. Tokita, "Transport phenomena in gel", *Gels,* vol. 2, no. 2, pp. 17-31, 2016. [http://dx.doi.org/10.3390/gels2020017]

[29] R. Trampel, J. Schiller, L. Naji, F. Stallmach, J. Kärger, and K. Arnold, "Self-diffusion of polymers in cartilage as studied by pulsed field gradient NMR", *Biophys. Chem.,* vol. 97, no. 2-3, pp. 251-260, 2002. [http://dx.doi.org/10.1016/S0301-4622(02)00078-9] [PMID: 12050014]

[30] G. Fleischer, F. Rittig, and C. Konak, "Influence of associations on self-diffusion of styrene-methyl-methacrylate random copolymers in semidilute acetone solution", *J. Polym. Sci., B, Polym. Phys.,* vol. 36, no. 16, pp. 2931-2939, 1998. [http://dx.doi.org/10.1002/(SICI)1099-0488(19981130)36:16<2931::AID-POLB8>3.0.CO;2-N]

[31] H. Walderhaug, and B. Nystrom, "Anomalous diffusion in an aqueous system of a poly(ethylene oxide)–poly(propylene oxide)–poly(ethylene oxide) triblock copolymer during gelation studied by pulsed field gradient NMR", *J. Phys. Chem. B,* vol. 101, no. 9, pp. 1524-1528, 1997. [http://dx.doi.org/10.1021/jp962197g]

[32] M. Nydén, and O. Söderman, "An NMR self-diffusion investigation of aggregation phenomena in solutions of ethyl(hydroxyethyl)cellulose", *Macromolecules,* vol. 31, no. 15, pp. 4990-5002, 1998. [http://dx.doi.org/10.1021/ma971472+] [PMID: 9680439]

[33] K. Thuresson, B. Nystrom, G. Wang, and B. Lindman, "Effect of surfactant on structural and thermodynamic properties of aqueous-solutions of hydrophobically-modified ethyl(hydroxy-ethyl)cellulose", *Langmuir,* vol. 11, no. 10, pp. 3730-3736, 1995. [http://dx.doi.org/10.1021/la00010a024]

[34] G. Fleischer, C. Konak, A. Puhlmann, F. Rittig, and J. Karger, "Dynamics of a triblock copolymer in a selective solvent for the middle block investigated using pulsed field gradient NMR", *Macromolecules,* vol. 33, no. 19, pp. 7066-7071, 2000. [http://dx.doi.org/10.1021/ma0003497]

[35] W.E. Baille, and X.X. Zhu, "Study of self-diffusion of hyperbranched polyglycidols in poly(vinyl alcohol) solutions and gels by pulsed-field gradient NMR spectroscopy", *Macromolecules,* vol. 37, no. 23, pp. 8569-8576, 2004. [http://dx.doi.org/10.1021/ma049588a]

[36] M. Appel, G. Fleischer, J. Karger, F. Fujara, and I. Chang, "Anomalous segment diffusion in polymer melts", *Macromolecules,* vol. 27, pp. 4274-4277, 1994. [http://dx.doi.org/10.1021/ma00093a031]

[37] V.V. Rodin, T. Cosgrove, and D. Qiu, "Inter and intra-molecular association in hydrophobic/ hydrophilic random copolymers", In: *ISIS Experimental Report (Rutherford Appleton Laboratory), RB No 20002*, 2005.

[38] V.V. Rodin, T. Cosgrove, U. Keiderling, and M. Murray, "Random copolymer association in solutions", In: *BENSC Experimental Report CHE-04-1274, Germany*, 2006.

[39] T. Cosgrove, V.V. Rodin, M. Murray, and Y. Espidel, NMR Studies of the structure of copolymer solutions and films.*Bristol Colloid Centre annual review*, 2005/2006, p. 27.

CHAPTER 8

Copolymer Films Swollen in Water: NMR Relaxation and PFG NMR Techniques in Studying Polymer-Water Interactions

Abstract: NMR relaxation and NMR diffusion techniques have been applied to investigate the films from the copolymers of methacrylic acid and butyl methacrylate after their swelling in water. ^{1}H NMR spectra and spin-spin relaxation times have been analysed to estimate the contributions of water protons and the protons of polymer matrix. NMR diffusion study in the films swollen in water discovered a dependence of the echo-attenuation signal on the diffusion time. The data obtained showed a restricted diffusion of water trapped in film pores. The results have been analysed taking into account the known physical models of water diffusion in materials. Applying the Tanner's approach to the data on water self-diffusion in the copolymer films, the pore size and permeability have been estimated. The increase of the water immersion time resulted in an additional water uptake and increasing molecular mobility of water.

Keywords: Apparent diffusion coefficient D_{app}, Copolymer films, Correlation times, Fourier transform (FT), Free induction decay (FID), NMR spectra, Permeability, Pulsed Field Gradient (PFG), Restricted diffusion, Spin-lattice (longitudinal) relaxation time T_1, Spin-spin (transverse) relaxation time T_2, Stimulated echo (STE), Water.

POLYMER FILMS AND WATER

Water is crucial in changing physical and chemical properties of different materials including polymers [1 - 3]. The questions of water interaction with polymer materials have been the object of many studies during long time [3 - 14]. In spite of extensive research and obtained results, the state of water in copolymers and polymer compounds so far remains not clear. The effects of water on polymer composites are very important in many practical applications [1 - 5, 8 - 11, 13 - 15].

Different methods including NMR techniques have been applied effectively in studying interactions of water with polymers [1, 3 - 12, 15 - 20]. Proton NMR relaxation times (T_1 and T_2) were used as informative parameters to examine hydration-dehydration effects in the polymers [3, 4, 8 - 12, 16 - 18]. In Ref. [17],

Victor V. Rodin

the NMR relaxation times were analysed as average values, *i.e.,* the signals were presented both from water protons inside the polymer as well as from exchangeable polymer protons. Those data indicated that the mobility of these protons varied by a factor of 10 or even more for the hydrogel contact lenses [17].

NMR relaxation techniques were used to study hydration properties of polymer materials and water transport [3, 4, 6, 7, 9 - 12]. On the base of NMR data, the properties of water are often described in the terms at least two different types of water: water strongly interacting with the polymer, and water which practically do not interact with the polymer matrix [14, 18, 20]. The bound water can be divided further on tightly and loosely bound waters. If water molecules form hydrogen bonds with the polar groups of the polymers, then these molecules can be considered as tightly bound water [14]. Thus, different characteristics of water (levels of mobility) in polymers could be described within concept of tightly bound and loosely bound water. In particular, the water strongly interacting with the polymer does not participate in water transport or solute transport in hydrated materials [10 - 13].

In Ref. [4], it was shown how NMR relaxation method characterises the water environments inside various polymer films with the same polymer composition. The data obtained by Carr-Purcell-Meiboom-Gill (CPMG) pulse sequence on the films swollen in water resulted in the T_2 relaxation time distributions with three T_2 peaks. The pulsed field gradient (PFG) NMR techniques were applied also to investigate a diffusivity of water in the polymers [6, 8, 11, 12, 15, 18]. PFG is well established NMR technique to study the diffusion constants of solvent and macromolecules in polymer/copolymer solutions [21 - 24]. If the liquid/water is trapped in the pores of materials, then the models of restricted diffusion can describe its diffusivity. The use of published approaches can result in the estimates of restricted zones [25 - 27].

The process of film formation (*e.g.,* drying a polymer solution on the solid surface) can affect the final properties of a film as well as a film ability to water uptake [1, 2, 4, 8, 9]. During drying the polymer from the solution, the solvent evaporates and the initial solution state transforms into the aggregated state of homogenous film. The process of the film formation still has the opened questions. The ability to predict the drying behaviour of a polymer mixture is of great interest to the practice applications of polymer films. A technology of the film formulation is very important in industry of copolymer materials [1, 2].

In the case of the film formation from solutions of random copolymers with different numbers of hydrophilic monomer units, a hydrophobic/hydrophilic ratio in the chains of the polymers would be a major factor influencing on the solvent removal in drying process [9]. The studies of the films casted from the polymer solutions of different hydrophobicity are very important in developing new technologies of covers to prevent or regulate water transport into materials and preparations.

This chapter discusses how the methods of NMR spectroscopy, NMR relaxation and NMR diffusion produce valuable information about the copolymer films formed from copolymer solutions of methacrylic acid (MAA) and butyl methacrylate (BMA) with different compositions and their interaction with water. NMR relaxation and PFG NMR tools demonstrated how the studies of films swollen in water can detail the film microstructure and properties. The role of residual solvent in water uptake is evaluated. The diffusion constants of water trapped in the restricted zones can be used in the estimates of the pore sizes.

DRIED AND WET POLYMER FILMS: ^{1}H NMR SPECTRA AND T_2 STUDIES OF WATER-POLYMER INTERACTIONS

Films in Experimental Conditions

For preparation of the films, the copolymer (32%) solutions of BMA-MAA in isopropanol (IPA) have been used. The molar ratio between BMA and MAA in the BMA-MAA copolymer films was variable [6]. The details of drying solution on PTFE or PET surface can be found in Refs. [6, 9]. Duration of the film drying was controlled by NMR analysis from day to day. The liquid-like signals of IPA disappeared and only a broad resonance solid-like signal could be recorded in ^{1}H NMR spectrum. A gravimetric analysis have been used to control the water content after immersion of the films into water environment [5].The MAA & BMA monomers and the pictures of the films are presented in Fig. (**1**).

Dried and Wet Polymer Films: NMR Spectra, FIDs and NMR relaxation Times

^{1}H NMR spectroscopy was used to analyse the samples of the films upon drying [9]. The film samples at different stages of drying were dissolved in deuterated chloroform. Fig. (**2**) shows several ^{1}H NMR spectra recorded on these solutions at different time of drying the polyBMA films.

The solid-state NMR spectroscopy (^{13}C CP/MAS) has been applied to characterise the composition of dried copolymers [6]. The ^{13}C CP/MAS spectra showed following chemical shifts for the groups: –**C**O ~177–185 ppm, –O**C**H$_2$ ~65 ppm, –**C**– ~55 ppm, –C–**C**H$_2$ –C– (main chain) ~45 ppm, –**C**H$_2$–**C**H$_2$– ~30 ppm, **C**H$_3$–CH$_2$– (CH$_3$ in –OCH$_2$ –**C**H$_2$–CH$_2$–**C**H$_3$) ~20 ppm, **C**H$_3$–C– ~15 ppm.

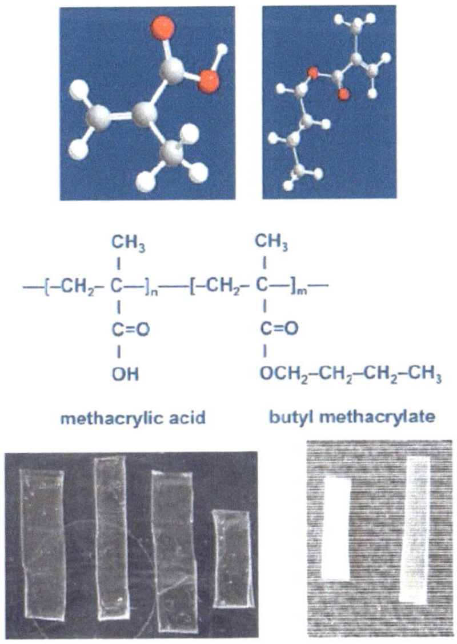

Fig. (1). *Top*: The monomers of MAA and BMA. *Bottom*: The pictures of BMA-MAA 100/0 (*left*) and BMA-MAA 90/10 (*right*) copolymer films dried during ~6 days. One of the BMA-MAA 90/10 copolymer films was swollen with water (*white*). The thicknesses h of the studied films were in the range of 0.7-1.1 mm. Dry films are transparent. After keeping in water, the films lose the optical transparency. The effect is called water whitening. NMR relaxation and optical transmission studies of the films immersed in water revealed [4] that some of the sorbed water is contained in the "bubbles" that scatter light.

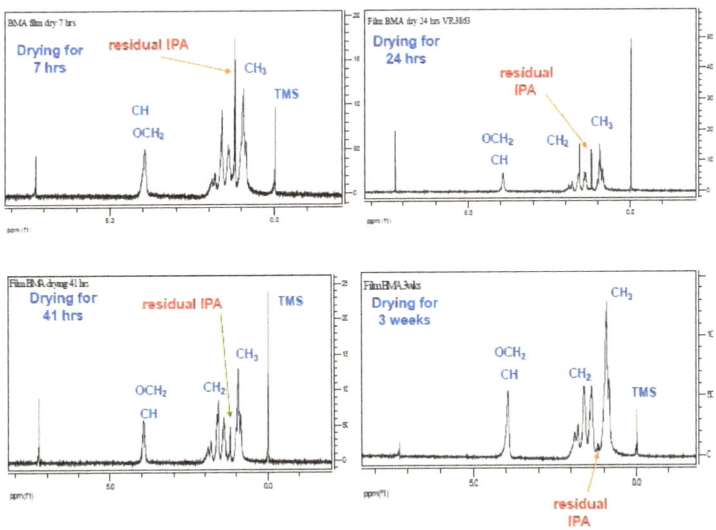

Fig. (2). ^{1}H NMR spectra (the *Bruker* DSX 300 NMR spectrometer, magnetic field 7.046 T) of the BMA-MAA 100/0 copolymer films dried on open air for 7, 24, 41 hrs and 3 weeks and then dissolved in $CDCl_3$. Reference is TMS, T=298 K. The residual signal of CH_3 group (~1.1 ppm) of IPA is still effectively present in the films dried for less than 2 days. Very small IPA signals were observable also in the films dried for about 3 weeks.

Fig. (**3**) presents two examples of ^{13}C CP/MAS NMR spectra recorded for dry copolymers BMA-MAA 100/0 (top) and 60/40 (bottom).

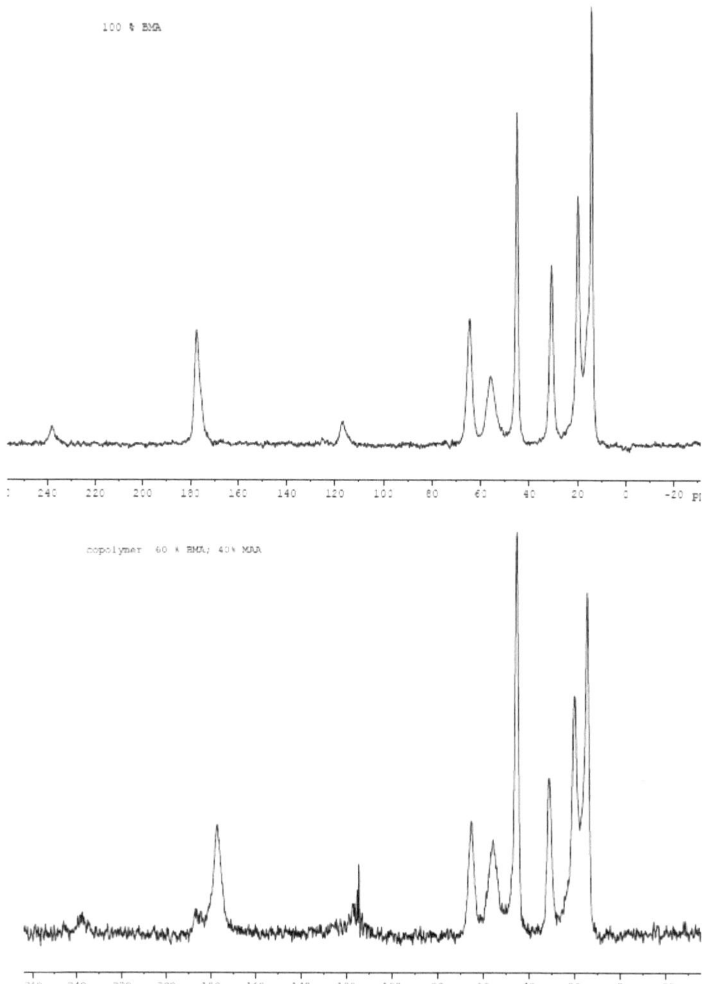

Fig. (3). ^{13}C CP/MAS NMR spectra of the dried copolymers.
top: BMA-MAA 100/0; *bottom*: BMA-MAA 60/40. Frequency is 75.4 MHz. T = 298 K.

Fig. (**4**) shows the ^{1}H NMR spectra of two dried copolymer films with different BMA-MAA composition (100/0 and 50/50). The shapes of the spectra were different for the films with different BMA content. The properties of the polymer films were depending on the residual solvent in the material. The changes of drying conditions and polymer composition resulted in different level of residual solvent in the copolymer films under study. For example, an increase of the temperature accelerated the process of drying, and the ^{1}H NMR spectra showed broad spectral peak then. Fig. (**5**) shows the ^{1}H NMR spectra for the BMA-MAA 60/40 copolymer films dried at the different temperature and time. Long time drying (4 days) at room temperature resulted in the formation of the film showing the ^{1}H NMR spectrum of relatively mobile protons. The broad ^{1}H NMR spectrum is registered for the BMA-MAA 60/40 copolymer film after drying for 1.25 days at 45°C testifying very low level of residual solvent in dried material.

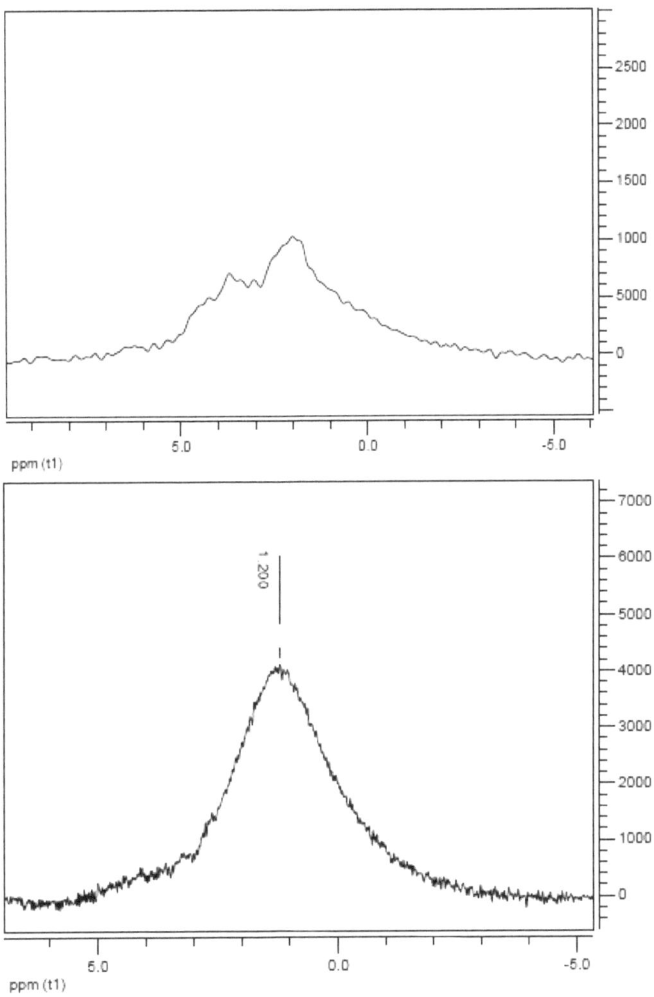

Fig. (4). The ^{1}H NMR spectra (the *Bruker* 300 MHz) of the BMA-MAA 100/00 (*top*) and the BMA-MAA 50/50 (*bottom*) copolymer films dried for 5 days on open air. T=298 K.

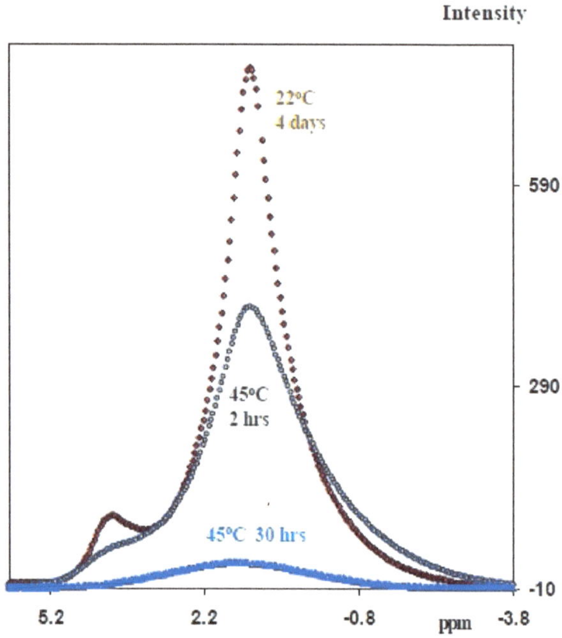

Fig. (5). The ^{1}H NMR spectra (the *Bruker* 300 MHz) of the BMA-MAA 60/40 copolymer films at different drying conditions: 22°C, 4 days; 45°C, 2 hrs; 45°C, 30 hrs (*blue*). The NMR measurements were made at room temperature.

If not fully dried films are immersed in water, the effect could be different for the films which had different drying treatment, *e.g.*, drying in air or drying in oven with heating. Water uptake in the films dried in these different conditions resulted in records of ^{1}H NMR spectra presented in Fig. (**6**). The findings show that the water uptake in the film after air drying and additional heating in oven (45°C) resulted in weakly intensive peak (~4.7 ppm) in the ^{1}H NMR spectrum. The water uptake in the film dried on open air without additional heating gave intensive water peak (Fig. **6**). Thus, after drying the BMA-MAA 60/40 copolymer films on the open air without heating, significant amount of residual IPA remains in the film. Such a film had effective water uptake at keeping that in water. However, the amount of residual solvent in the film dried in oven (45°C) was already less than that in the case of drying on open air. That film showed low water uptake. For the BMA-MAA copolymer films obtained at different composition and drying conditions, the FIDs were characterised by different shapes [6, 8, 9]. The resonance peaks in ^{1}H NMR spectra of these films are also described by different line shapes. In most cases, the peaks are *Lorentzian* and *Gaussian* line shapes or combination of both (*Gaussian* broadening of liquid-like signal) [9, 28].

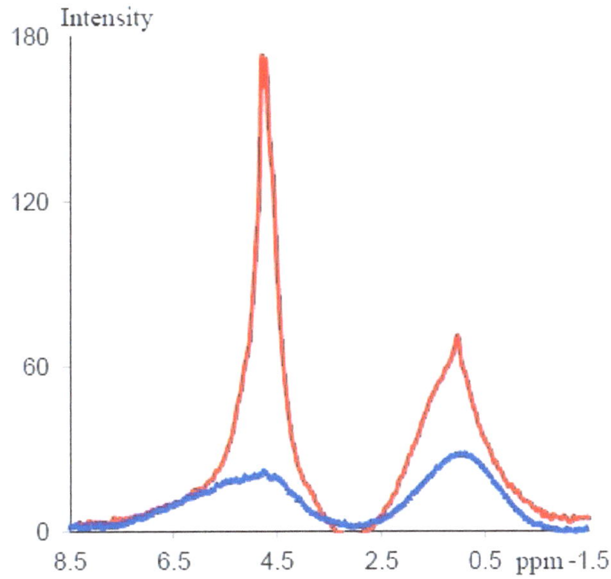

Fig. (6). The ^{1}H NMR spectra (the *Bruker* 300 MHz) after water storage (2 days) of the BMA-MAA 60/40 copolymer films dried at different conditions: open air, 22°C (*red*) and oven, 45°C (*blue*). The NMR measurements were made at room temperature.

When the spectrum of the protons is described by *Lorentzian,* the time-domain magnetization M_L is expressed by the exponential signal:

$$M_L(t) = M_0 e^{-\pi\omega_L t} \quad \textbf{(1)}$$

When the spectrum is *Gaussian* line shape, the time-domain signal M_G is presented according to:

$$M_G(t) = M_0 e^{-\left(\frac{\pi\omega_G t}{2(\ln 2)^{1/2}}\right)^2} \quad \textbf{(2)}$$

The *Lorentzian* and *Gaussian* widths of the spectral lines (at half height of the signals) in the frequency domain are titled ω_L and ω_G, respectively.

Ref. [6] presented time-domain NMR signals for the BMA-MAA 100/0 and 60/40 copolymer films. The FIDs of these films were considered as the sum of the solid-like decaying part (fast relaxing component) and liquid-like contribution (slower decaying component) according to published approaches [6, 28, 29]. The *sinc* function and *Gaussian* broadening were combined to fit the fast relaxing component of the magnetization decay [29, 30]. The *Voigt* function described the slow relaxing component [29 - 31].

Depending on the drying conditions and composition of the BMA-MAA copolymer films, the fits of the slow relaxing components in the FIDs often used two exponential functions instead *Voight* function. Based on these fits, the range of correlation times τ_c was estimated from 0.6×10^{-6} s to 3.0×10^{-6} s for the slowest relaxing component of the FIDs in the films upon drying [6].

In the case of distribution $S(\tau)$ of molecular correlation times for segmental motion in polymers, the data on transverse magnetization $M(t)$ are connected with $S(\tau)$ by the expression (3) [28, 32]:

$$M(t) = \int_0^\infty S(\tau)\exp(-t/T_2)d\tau \quad \textbf{(3)}$$

Studying the films with thicknesses from 0.4 to 0.8 mm showed that correlation times $\tau_c \sim 1\times10^{-6}$ c characterised the proton fraction with population $\leq 3\%$ of all measurable protons. In the calculations, the relationship between τ_c, second moment, spectrometer frequency and the T_2 of the protons of polymer (with distribution from eq. (3)) have been used [30, 32, 33]. For the series of dried films with different BMA-MAA molar content, the measurements resulted in average second moment as $(10 \pm 2) \times 10^9$ s^{-2} [6].

At the study of water uptake in the BMA-MAA copolymer films with variable composition, the *Carr-Purcell-Meiboom-Gill* (CPMG) sequence, $90°_x - (t_{cp} - 180°_y - t_{cp})_n$, could be applied to measure the T_2 and to quantify the populations of water in different environments [6, 8, 29, 31]. When the bulk water interacts with inner surface of the pores in films, the T_2 values became shorter. The size of the pores defines inner surface area interacting with porous water. Water in physical environments of different sizes is characterized by different transverse relaxation times T_2. In porous media with distribution of compartment sizes, the CPMG measurement (with the n spin-echoes) could be analysed by the eq. (3) [30, 33], where the time after the 90° pulse is $t = 2nt_{cp}$. The pulse interval between 90° and 180° in CPMG experiment is t_{cp}. The distribution S in eq. (3) should be considered then as T_2 distribution function. This could be defined using the inverse Laplace transformation [4, 31, 34, 35]. Studying porous systems, NMR methods measure the pore size distribution in a saturated material. The spin-spin relaxation time T_2 is inversely proportional to the surface area-to-volume ratio of the pore [35 - 38]. Thus, radius distribution can be analysed on the base of the $S(T_2)$ data [36].

The study of water uptake in the dried pBMA polymer film showed [6] that dried film (before keeping the sample in water) had the broad spectral line with two weak humps at ~1-3 and ~4-5 ppm. Dried pBMA film had some amount of residual solvent indeed, which was not observable within broad residual line because of low mobility of the protons (Fig. **4**). After keeping the film sample in water, *e.g.,* for 18 hrs, two signals (informing about mobile protons inside the film) appeared over the humps [6]. Fig. (**7**) shows the ^{1}H NMR spectrum for the pBMA film kept in water for longer time (6 days). The peaks are already more intensive than in the case of short water exposure [6].

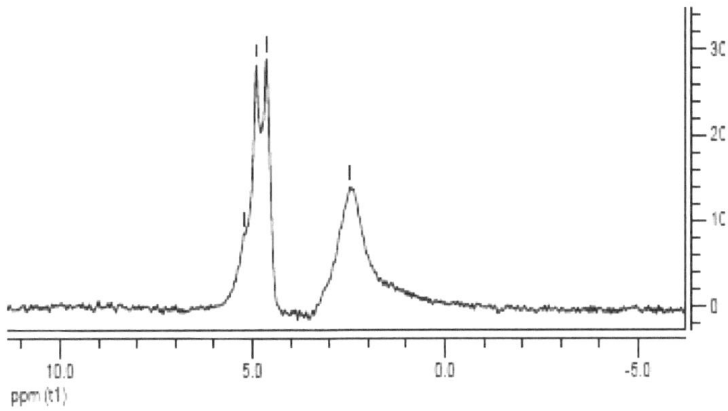

Fig. (7). ^{1}H NMR spectrum (*The Bruker* 300 MHz) of pure pBMA film after 6 days storage in H_2O. T=298 K. *h*=1 mm.

During swelling the film in water, the molecules of residual IPA exchanged with water molecules, and the mobility of IPA in the film increased sufficiently that its CH_3 groups could be detected in ^{1}H NMR spectra. The proton resonance of CH_3 group of IPA is usually observable at ~1.1 ppm in the ^{1}H NMR spectra of the IPA-solutions / BMA-MAA copolymer films dissolved in deuterated solvents [6, 9, 21, 23] (Fig. **2**). The ^{1}H peak at ~ 4-5 ppm in the spectra of films swollen in water (Figs. **6**, **7**) was connected with OH resonance. This was also confirmed in exchange experiments with H_2O/D_2O [6].

The change of hydrophilic/hydrophobic ratio in the copolymer resulted in changing the water uptake in the films. For example, the BMA-MAA 60/40 copolymer film was more sensitive to water: the intensity of OH signal in this film was larger than that in 100/0 films. The OH groups of the IPA interact with the MAA units. Water also interacts well with OH groups of MAA. When one increases the number of MAA units in the film, the number of interacting solvent molecules per polymer chain must increase. That is why after 6 hrs in water the BMA-MAA 60/40 film had more intensive mobile signal (the slowest relaxing T_2 component had a population 60% of all measurable protons). The BMA-MAA

100/0 film after 6 hrs of water action resulted in less intensity of slowest relaxation component in T_2 experiment (~30%) [6].

Water and residual isopropanol in the film are in the exchange regime. Due to the exchange, IPA released into the surrounding volume of water during the storage of the film in water. Fig. (**8**) shows the ^{1}H NMR spectrum for pure isopropanol (*up*) and that for the sample of water in which the polymer film was stored (*bottom*). Despite the intense water signal in the region of ~4.8 ppm, the CH_3 (~1.1 ppm) and CH (~4 ppm) signals of the isopropanol are also recorded, thus confirming that part of the residual isopropanol left the film during keeping the film in water. Additional experiments with the storage of the films in D_2O confirmed the escape of IPA from the film into water/D_2O (Fig. **9**).

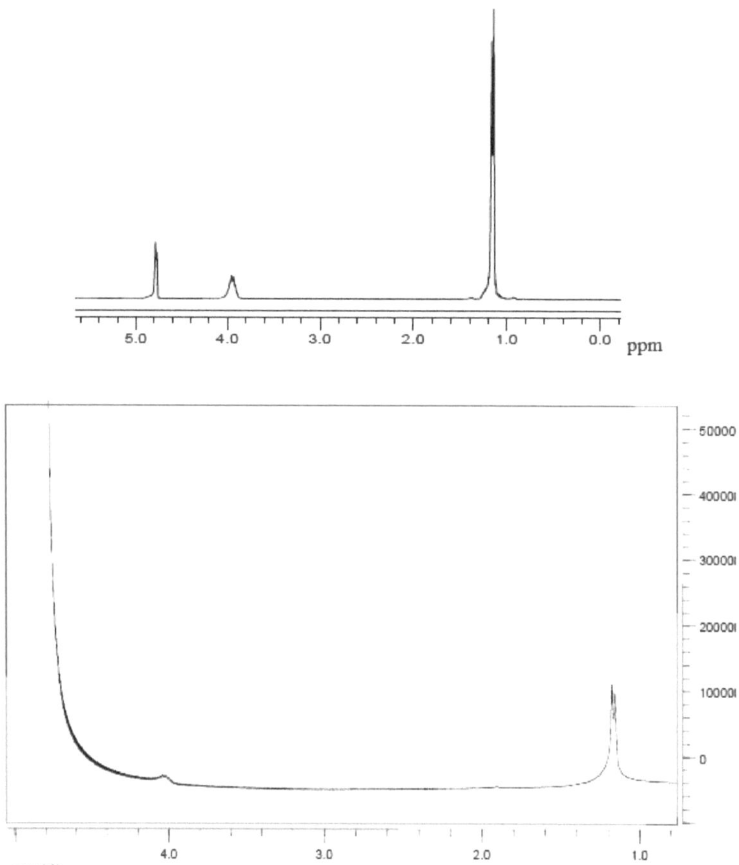

Fig. (8). *up*: ^{1}H NMR spectrum (*The Bruker* 300 MHz) of IPA.
bottom: ^{1}H NMR spectrum (*The Bruker* 300 MHz) of H_2O in which the BMA-MAA 10/90 copolymer film was kept for 10 days. T=298 K. The IPA signals observed in water volume were ~1.1 ppm (CH_3 group) and ~4.0 ppm (CH group).

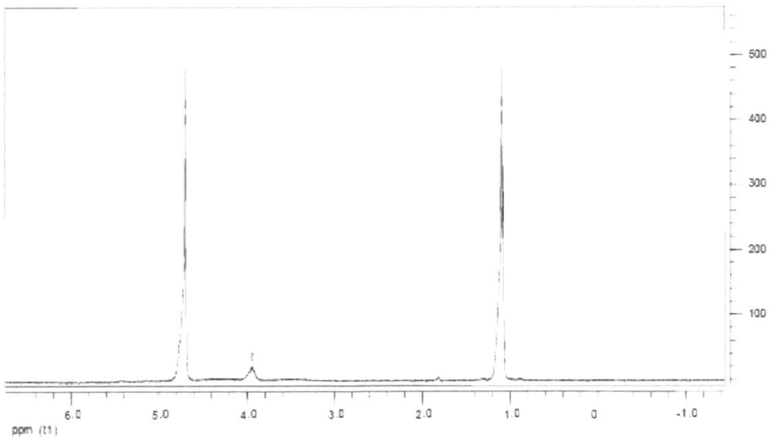

Fig. (9). ^{1}H NMR spectrum of D_2O in which the BMA-MAA 10/90 copolymer film was stored for 5 hrs. The film was casted from copolymer-IPA solution during 6 days drying on open air. After the sample storage in D_2O, the film was removed and then D_2O was analysed.
IPA signals in D_2O sample: ~1.1 ppm (CH_3 group) and ~4.0 ppm (CH group). A signal of residual H_2O in D_2O (OH group) and that of OH group (IPA) were registered as one signal at ~4.7 ppm.

Fig. (**10**) (up) shows the ^{1}H NMR spectrum of the BMA-MAA 50/50 film after water exposure (13 hrs). In addition to this, the spectrum characterising only water and IPA in this film could be observed as a difference spectrum between the spectrum of swollen film and that of dried film. Fig. (**10**) (bottom) presents such a difference spectrum informing about the water (OH) peak (~5-6 ppm) and about the peak of IPA ($(CH_3)_2$ groups of IPA). With this approach, the signal of non-exchangeable protons of IPA (presented in dry film after removing water and IPA from the film) can be eliminated from observation [8]. Thus, the findings show that the signal of CH_3 group of IPA is less intensive in difference spectrum than that in swollen film (Fig. **10**).

The role of hydrophobic domains of the copolymers and the effect of water uptake on the polymer properties and plasticization were studied in the publications [19, 39]. According to these data, the monomer nature as well as hydrophobic/hydrophilic molar ratio influence the amount of water absorbed in the polymers and water diffusion in the copolymers. Applying the MRI to the copolymers of hydroxyethyl methacrylate (HEMA) with tetrahydrofurfuryl methacrylate (THFMA), the authors studied water diffusion inside the polymer matrices with different monomer composition [39].

A state of water in the polymer films can be characterised by correlation times τ_c calculated with the aid of eq. (12) from Ref. [6] with preliminary estimation of bound water fraction η (ratio of water molecules number connected to the polymer surface to all molecules of water). For water molecules that entered the polymer film BMA-MAA 60/40 after it was held for 6 hours in water, the correlation times were in the range of (3.8 ± 0.3)×10^{-7}s. In the BMA-MAA 100/0 copolymer films after the same duration of storage in water, the τ_c values were (4.1 ± 0.3)×10^{-7}s [6]. An increase of the storage time of the films in water or varying content of BMA (MAA) monomers resulted in the change of water characteristics in copolymer matrix [4, 6, 9].

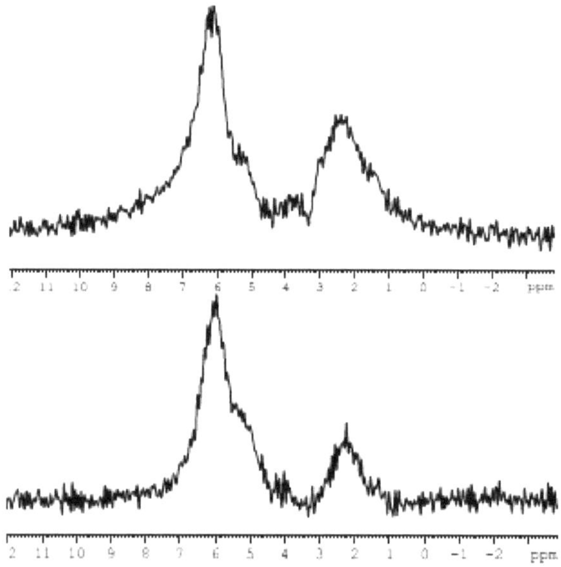

Fig. (10). ^{1}H NMR spectrum for the BMA-MMA 50/50 copolymer film swollen in water for 13 hrs (up). The difference spectrum (*bottom*) is obtained after subtraction of the spectrum of dried film from the (up) spectrum. T=298 K. The frequency is 300 MHz.

The NMR signals become more intensive (Figs. **6**, **7**) when additional amount of water penetrates into the film. Then, the rate of transverse relaxation for water inside the films would be much slower. The study of water uptake in the copolymer films (when the BMA/MAA 60/40 and 80/20 copolymer films were placed in water for 12 hrs) by CPMG pulse sequence resulted in two exponential components in the fits of CPMG decays. The fast relaxing component was associated with the correlation times τ_c in the range from 0.42×10^{-8} s to 0.74×10^{-8} s, whereas for slow relaxing component, the range of τ_c between 1.2×10^{-10} s and 1.6×10^{-10} s has been calculated [6]. This implies that the slow relaxing component in absorbed water is responsible for the water molecules which are not directly bound to the surface of pores. Fast relaxing component characterizes the molecules of water in close interaction with the polymer, *i.e.*, the protons in the first and the second hydration layers in pores of polymer network.. The data on the BMA-MAA 90/10 copolymer film swollen in water (at humidity level of 8.4% by weight) resulted in two components (measured by CPMG) with transverse relaxation time $T_2' = 112$ ms (83%, slow component) and $T_2'' = 41$ ms (17%, fast relaxing component). Then, it is possible to suggest that the slow relaxing component (*i.e.*, T_2') is responsible for the protons of water in the large holes or voids. The component with small T_2'' value explains then water dynamics in the film pores of small size. In the case of exchange between water fractions, more complicated description of relaxation components could be considered [4, 10, 12, 18].

WATER SELF-DIFFUSION IN THE COPOLYMER FILMS

Fig. (**7**, chapter 1) shows the PFG stimulated-echo experiment (with three 90° pulses) that was used for NMR diffusion studies in sorbed water. The attenuation of echo in PFG experiment is presented by following eq. (4) [18, 21 - 23, 27, 29]:

$$I(b)/I(0) = \exp(-2\tau_1/T_2 - \tau_2/T_1)\exp(-bD) \quad \textbf{(4)}$$

Here, $I(0)$ is the echo intensity without gradient, $I(b)$ is an amplitude of echo in presence of gradient. Δ is the time between the gradient pulses, γ is the proton *gyromagnetic ratio* equaled to 267.51×10^6 rad$\times$T$^{-1}\times$s^{-1}, b is the gradient factor $(\gamma G\delta)^2(\Delta-\delta/3)$, τ_1 is the time between the 1st and 2nd $\pi/2$ pulses, τ_2 is the time between the 2nd and 3rd RF pulses. The chapters 3, 5, 7 considered the dependence of echo intensities for free and restricted diffusion.

Fig. (**11**) shows how self-diffusion coefficient of water in the BMA-MAA 90/10 copolymer film is depending on diffusion time Δ. Earlier in Refs. [21 - 23, 27, 40], a dependence of apparent diffusion coefficient $D_{app}=f(\Delta)$ was detailed for the quasi-restriction case. In accordance with those approaches, it was found that water diffusivity in BMA-MAA 90/10 copolymer films obeys to the law $D_{app}\sim\Delta^{k-1}$ where k-value is less than 1 [27, 40]. Thus, the water diffusion in studied copolymer films is quasi-restricted. To understand the details of water diffusion in the film structure further, it is possible to consider a continuous medium which has an arbitrary permeability p and barrier spacing a. According to publications [6, 25, 41 - 43], the echo-attenuation function for diffusion within reflecting spherical boundaries (at long time-scale limit $\Delta >> a^2/2D$) can be considered as follows:

$$E(G,\delta) = \frac{9[\gamma G\delta a\cdot\cos(\gamma G\delta a) - \sin(\gamma G\delta a)]^2}{(\gamma G\delta a)^6} \quad \textbf{(5)}$$

An apparent diffusion coefficient in the system with spherical boundary is presented as $a^2/5\Delta$ [41 - 44]. Echo-attenuation function $E(G,\delta)$ is expressed then as follows: $E(G, \delta) = \exp[-(\gamma G\delta a)^2/5]$ [6, 41, 43]. Film structure results in the water diffusion restriction and defines the Δ dependency of D_{app} for water in swollen film (Fig. **11**). It was suggested to consider the apparent diffusion coefficient for the range of small Δ as the free diffusion coefficient D_0 [25, 26, 42]. At the increasing Δ, the apparent diffusion coefficient decreases. At very long diffusion times, D_{app} approaches asymptotic value D_{asym}. This asymptotic diffusion coefficient is connected with barrier spacing a, the free diffusion coefficient D_0 and permeability p by following eq. (6) [26, 29, 41, 42, 44]:

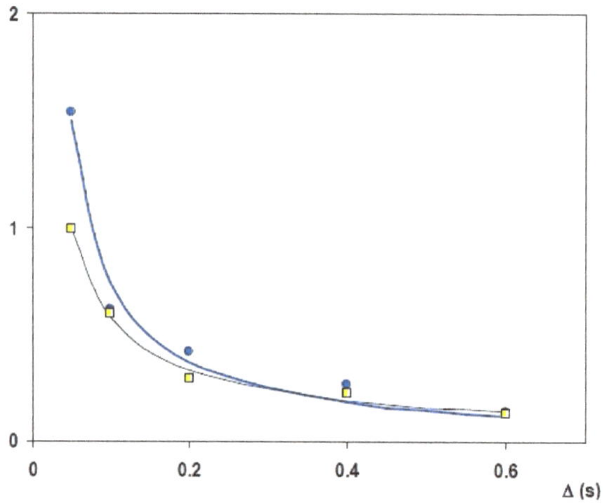

Fig. (11). The water diffusion coefficients in the BMA-MAA 90/10 films with water content of 7.4% (*squares*, 50 days in water) and 8% (*circles*, 51 days in water) *vs* diffusion time with a fit to a restricted diffusion model: $D = Q\Delta^{k-1}$ (Q is a constant, $k < 1$). *Bruker* DSX 300 NMR spectrometer (frequency for ^{1}H is 300 MHz) had diff30 probe for PFG NMR experiments. G_{max}=10 T/m. δ=1 ms. T=298 K.

$$D_{asym} = \frac{D_0 ap}{D_0 + ap} \qquad (6)$$

Thus, from experiment it is possible to estimate D_{asym}, D_0 and further to know the product ap. For analysis of ap it is possible to use intermediate parts of the experimental dependency of D_{app} on diffusion time [25, 26]. The approach suggested by *Tanner* [26] considers relative apparent diffusion coefficient D_{app}/D_0 *vs* reduced diffusion time $D_0 t/a^2$. D_{app}/D_0 approaches 1 in the limit of zero time.

Tanner used the point in the dependency of D_{app} on diffusion time where D_{app} is equal to its average value to read off diffusion time $t_{1/2}$ and to calculate further restriction size a, and permeability p. On the base of this procedure, the data of Fig. (**11**) resulted in estimates of pore size $a \approx 0.61$ μm with permeability $p \approx 2.6 \times 10^{-5}$ cm/s (water content ~7.4%) and comparable restriction size $a \approx 0.54$ μm with permeability $p \approx 2.77 \times 10^{-5}$ cm/s for the BMA-MAA 90/10 copolymer film with water content of 8%.

The analysis of data for the polymer film casted from IPA solution of polyBMA (water storage for 18 hours) showed also restricted diffusion according to dependency of $D_{app} \sim \Delta^{k-1}$ where k is less than 1 [6, 8, 29, 40]. The estimate made with the same approach [26] resulted in a $\approx$ 0.85 μm. Some difficulties in applying restriction diffusion theory to discover the actual structure of porous media with various barriers have been discussed in Refs. [25, 26, 40 - 43, 45 - 48]. As to polymer film, the surface of pores causes the restrictions. There is also a problem with an estimation of D_{asym}. The values of diffusion coefficients did not line up in an apparent constant even at very long diffusion time [40 - 43]. Several different diffusion coefficients (or a distribution of D values) can exist in porous materials. An exchange between different components should be taken into account then.

The value of D_{app} is depending on Δ. For the film samples with comparable water content, the results showed that the average sizes of restrictions can be between 0.54 and 1 μm [6]. The experimental CPMG decays fitted by the two exponentials could characterise formed small and large pores. The estimates of their sizes could be carried out then on the base of T_2 data [6, 36, 42, 49, 50]. For example, Ref. [49] considered water sorbed in the pores as the molecular probe and estimated the size of pores in moist coals using NMR relaxation data. The pores in the material under study filled with the pore water were slit-like [49]. Ref. [50] gives a connection between the spin-spin relaxation time T_2 in a spherical pore with size a, the surface relaxation time T_{2s}, and the thickness of the surface layer λ (that is considered to be ~3 Å) by the expression $T_2 = (T_{2s}/\lambda)\cdot(a/3)$. The constant $a/3$ depends on the shape of pores. For a cylindrical pore shape, this constant varies by $a/2$. One estimates a surface relaxation time circa between 60 and 100 μs [50].

Conversion of the T_2 relaxation data of the BMA-MAA 90/10 film into pore sizes (using literature approaches of estimates of pore sizes in various materials) results in estimates of the pores in copolymer films between 0.8 and 1.2 μm. Thus, the estimates of pore sizes from diffusion data on the films are very close to those from the relaxation data. The electron microscopy pictures of the copolymer films (CRYO-SEM data [6]) discovered also the spherical pores with the sizes distributed near most frequently measured values ~ 1 μm.

CONCLUDING REMARKS

It was shown how NMR diffusion and NMR relaxation methods produce a characterization of molecular mobility at casting the films from copolymer (butyl methacrylate and methacrylic acid) solutions and of water diffusion in the copolymer films at their storage in water. The increase of film storage time in water resulted in the additional water absorption and increased mobility of water molecules inside the films. At the variation of the hydrophilic monomer MAA fraction in the copolymers, it was possible to cast the films with different composition and to change the water uptake in the films.

NMR diffusion studies showed how apparent diffusion coefficient was depending on the time of diffusion. The data obtained testified about quasi-restricted self-diffusion of water in swollen copolymer films. The models of water diffusion in spherical boundaries and the size of the restriction zone have been analysed to fit the data obtained. The details of water diffusion in the films have been considered in connection with film microstructure. The walls of the pores with permeability *p* presented the barrier for diffusion of water molecules. The sizes of the pores in the films estimated by NMR diffusion, NMR relaxation, and electron microscopy were comparable.

ACKNOWLEDGEMENTS

The author thanks Terence Cosgrove and Martin Murrey.

REFERENCES

[1] Y. Reyes-Mercado, F.J. Rodríguez-Gómez, and Y. Duda, "Effect of the acrylic acid content on the permeability and water uptake of latex films", *Colloid Polym. Sci.*, vol. 286, no. 5, pp. 603-609, 2008. [http://dx.doi.org/10.1007/s00396-008-1838-6]

[2] A.Y. Koh, S. Mange, M. Bothe, R.J. Leyrer, and R.G. Gilbert, "The influence of copolymerization with methacrylic acid on poly (butyl acrylate) film properties", *Polymer,* vol. 47, no. 4, pp. 1559-1165, 2006. [http://dx.doi.org/10.1016/j.polymer.2005.12.053]

[3] P. McConville, M.K. Whittaker, and M. Pope, "Water and polymer mobility in hydrogel biomaterials quantified by ^{1}H NMR: A simple model describing both T_1 and T_2 relaxation", *Macromolecules,* vol. 35, no. 18, pp. 6961-6969, 2002. [http://dx.doi.org/10.1021/ma020539c]

[4] Y. Liu, A. Gajewicz, V.V. Rodin, W-J. Soer, J. Scheerder, G. Satgurunathan, P.J. McDonald, and J.L. Keddie, "Explanations for water whitening in secondary dispersion and emulsion polymer films", *J. Polym. Sci., B, Polym. Phys.*, vol. 54, no. 16, pp. 1658-1674, 2016. [http://dx.doi.org/10.1002/polb.24070]

[5] K. Krysztofiak, and A. Szyczewski, "Study of dehydration and water states in new and worn soft contact lens materials", *Opt. Appl.*, vol. 44, no. 2, pp. 237-250, 2014.

[6] V.V. Rodin, and T. Cosgrove, "Nuclear magnetic resonance study of water-polymer interactions and self-diffusion of water in polymer films", *OALJ: Chem. Mater. Sci,* vol. 3, no. 10, pp. 1-17, 2016.

[7] V.M. Bouznik, E.V. Morozov, and I.A. Avilova, "NMR applications for polymer composite materials moisture uptake investigation", *Appl. Magn. Reson.*, vol. 47, no. 3, pp. 321-334, 2016. [http://dx.doi.org/10.1007/s00723-015-0748-2]

[8] V.V. Rodin, T. Cosgrove, and M. Murrey, "Magnetic resonance imaging and pulsed field gradient NMR in study of structure of polymer solutions and films", *Proceedings of the 6th Colloquium on Mobile Magnetic Resonance,* 2006, pp. 32-33, Aachen, Germany.

[9] V.V. Rodin, T. Cosgrove, M. Murray, and R. Buscall, "Intermolecular interactions upon film formation from copolymer solutions", *Book of Abstracts XVIIth European Chemistry at Interfaces Conference,* 2005, p. 107, Loughborough.

[10] W.J. Soer, J. Scheerder, G. Satgurunathan, Y. Liu, V.V. Rodin, and J.L. Keddie, "Water transport through polymer films: comparison of solution, emulsion and secondary emulsion polymers", *International Conference (20-24 July 2014, Prague) 'Frontiers of Polymer Colloids: From Synthesis to Macro-Scale and Nano-Scale Applications',* 2014, p. 191.

[11] R. Barbieri, M. Quaglia, M. Delfini, and E. Brosio, "Investigation of water dynamic behaviour in poly(HEMA) and poly(HEMA-co-DHPMA) hydrogels by proton T_2 relaxation time and self-diffusion coefficient n.m.r. measurements", *Polymer,* vol. 39, no. 5, pp. 1059-1066, 1998. [http://dx.doi.org/10.1016/S0032-3861(97)00403-5]

[12] V. Baukh, H.P. Huinink, O.C. Adan, S.J. Erich, and L.G. van der Ven, "Water-polymer interaction during water uptake", *Macromolecules,* vol. 44, no. 12, pp. 4863-4871, 2011. [http://dx.doi.org/10.1021/ma102889u]

[13] M. Tanaka, T. Hayashi, and S. Morita, "The roles of water molecules at the biointerface of medical polymers", *Polym. J.,* vol. 45, no. 7, pp. 701-710, 2013. [http://dx.doi.org/10.1038/pj.2012.229]

[14] J. Munćan, I. Mileusnić, J. Šakota, J.Š. Rosić, A. Vasić-Milovanović, and L. Matija, "Water properties of soft contact lenses: a comparative near-infrared study of two hydrogel materials", *Intern. J. Polym. Sci.,* vol. 2016, 2016. Article ID 373791, 8 pages, 2016. doi:10.1155/2016/3737916

[15] C. Manetti, L. Casciani, and N. Pescosolido, "Diffusive contribution to permeation of hydrogel contact lenses: theoretical model and experimental evaluation by nuclear magnetic resonance techniques", *Polymer,* vol. 43, no. 1, pp. 87-92, 2002. [http://dx.doi.org/10.1016/S0032-3861(01)00559-6]

[16] A. Woźniak-Braszak, M.M. Kaźmierczak, M.K. Baranowski, K. Hołderna-Natkaniec, and K. Jurga, "The aging process of hydrogel contact lenses studied by ^{1}H NMR and DSC methods", *Eur. Polym. J.,* vol. 76, pp. 135-146, 2016. [http://dx.doi.org/10.1016/j.eurpolymj.2016.01.038]

[17] D.W. Larsen, J.W. Huff, and B.A. Holden, "Proton NMR relaxation in hydrogel contact lenses: correlation with in vivo lens dehydration data", *Curr. Eye Res.,* vol. 9, no. 7, pp. 697-706, 1990. [http://dx.doi.org/10.3109/02713689008999586] [PMID: 2209067]

[18] V.V. Rodin, B.V. Sakharov, and D.P. Knight, "Hydrated polymer matrices: NMR relaxation and pulsed gradient in molecular mobility studies", *Plast. Massy,* no. 2, pp. 10-16, 2000.

[19] P.Y. Ghi, D.J. Hill, and A.K. Whittaker, "NMR imaging of water sorption into poly(hydroxyethyl methacrylate-co-tetrahydrofurfuryl methacrylate)", *Biomacromolecules,* vol. 2, no. 2, pp. 504-510, 2001. [http://dx.doi.org/10.1021/bm000146q] [PMID: 11749213]

[20] F. Wang, R. Zhang, T. Chen, and P. Sun, " ^{2}H Solid-state NMR analysis of the dynamics and organization of water in hydrated chitosan", *Polymers (Basel),* vol. 8, no. 4, p. 149, 2016. [http://dx.doi.org/10.3390/polym8040149]

[21] T. Cosgrove, V.V. Rodin, M. Murray, and R. Buscall, "Self-diffusion in solutions of carboxylated acrylic polymers as studied by Pulsed Field Gradient NMR. 1. Solvent diffusion studies", *J. Polym. Res.,* vol. 14, no. 3, pp. 167-174, 2007. [http://dx.doi.org/10.1007/s10965-006-9087-1]

[22] R. Bai, P.J. Basser, R.M. Briber, and F. Horkay, "NMR water self–diffusion and relaxation studies on sodium polyacrylate solutions and gels in physiologic ionic solutions", *J. Appl. Polym. Sci.,* vol. 131, no. 6, 2014. [http://dx.doi.org/10.1002/app.40001] [PMID: 24409001]

[23] T. Cosgrove, V.V. Rodin, M. Murray, and R. Buscall, "Self-diffusion in solutions of carboxylated acrylic polymers as studied by pulsed field gradient NMR. 2. Diffusion of macromolecules", *J. Polym. Res.,* vol. 14, no. 3, pp. 175-180, 2007. [http://dx.doi.org/10.1007/s10965-006-9088-0]

[24] S. Viel, D. Capitani, L. Mannina, and A. Segre, "Diffusion-ordered NMR spectroscopy: a versatile tool for the molecular weight determination of uncharged polysaccharides", *Biomacromolecules,* vol. 4, no. 6, pp. 1843-1847, 2003. [http://dx.doi.org/10.1021/bm0342638] [PMID: 14606917]

[25] J.E. Tanner, "Transient diffusion in system partitioned by permeable barriers. Application to NMR measurements with a pulsed field gradient", *J. Chem. Phys.,* vol. 69, no. 4, pp. 1748-1754, 1978. [http://dx.doi.org/10.1063/1.436751]

[26] J.E. Tanner, "Self diffusion of water in frog muscle", *Biophys. J.,* vol. 28, no. 1, pp. 107-116, 1979. [http://dx.doi.org/10.1016/S0006-3495(79)85162-0] [PMID: 318065]

[27] V.V. Rodin, and V.A. Nikerov, "NMR relaxation and PFG NMR studies of water dynamics in oriented collagen fibres with different degree of cross-linking", *Current Tissue Eng.,* vol. 3, no. 1, pp. 47-61, 2014. [http://dx.doi.org/10.2174/2211542003666140626211652]

[28] V.V. Rodin, and G.M. Reznichenko, "Study of polymer structure by free induction decays", *Mater. Sci.,* no. 2, pp. 18-23, 2004.

[29] V.V. Rodin, "Methods of Magnetic Resonance in Studying Natural Biomaterials", In: *Encyclopedia of Physical Organic Chemistry,* Zerong Wang, Ed., Chapter 53. vol. 4, part 4. John Wiley & Sons, Inc., 2017, pp. 2861-2908. http://eu.wiley.com/WileyCDA/WileyTitle/productCd-1118470451.html

[30] V.V. Rodin, G.M. Reznichenko, and E.L. Vasina, "Properties of natural polymer fibers with low water content", *Mater. Sci.,* no. 1, pp. 34-42, 2004.

[31] V.V. Rodin, Magnetic resonance in studying cells, biotechnology dispersions, fibers and collagen based tissues for biomedical engineering. *Biological, Physical and Technical Basics of Cell Engineering.* G.M. Artmann, A. Artmann, A.A.Zhubanova, I.Digel, Eds., Springer: Singapore, pp.339-363, 2018. ISBN 978-981-10-7903-0, DOI:10.1007/978-981-10-7904-7.

[32] V.S. Derinovskii, Y.R. Yarda, I.N. Zakirov, and M.P. Myasnikova, "Molecular motion and nuclear magnetic relaxation in filled poly(vinyl acetate)", *Polymer Sci. USSR,* vol. 32, no. 4, pp. 776-782, 1990. [http://dx.doi.org/10.1016/0032-3950(90)90403-S]

[33] V.V. Rodin, G.M. Reznichenko, and E.L. Vasina, "Properties of natural silk fibers: deformation studies and NMR data", *Biophysics,* vol. 49, no. 6, pp. 918-926, 2004.

[34] P. Berman, O. Levi, Y. Parmet, M. Saunders, and Z. Wiesman, "Laplace inversion of low-resolution NMR relaxometry data using sparse representation methods. *Concepts.Magn Reson* ", *Part A,* vol. 42, no. 3, pp. 72-88, 2013.

[35] Y.-Q. Song, "Focus on the physics of magnetic resonance on porous media", *New J. Phys,* vol. 14, p. 055017, 2012. [http://dx.doi.org/10.1088/1367-2630/14/5/055017]

[36] A. Valori, *"Characterization of cementitious materials by ^1H NMR"*, PhD Thesis: University of Surrey, 2010.

[37] V.V. Rodin, A. Valori, and P.J. McDonald, "Pore-water interactions in hydrated cement pastes by NMR", *Proceedings of the 9th international Bologna conference on Magnetic Resonance in Porous Media* 13-17 July 2008, Cambridge, MA, USA, AIP Conference Proceedings.

[38] A. Valori, V.V. Rodin, and P.J. McDonald, "On the interpretation of ^1H 2-dimensional NMR

relaxation exchange spectra in cements: is there exchange between pores with two characteristic sizes or Fe^{3+} concentrations?", *Cement Concr. Res.*, vol. 40, no. 9, pp. 1375-1377, 2010. [http://dx.doi.org/10.1016/j.cemconres.2010.03.022]

[39] P.Y. Ghi, D.J. Hill, D. Maillet, and A.K. Whittaker, "NMR imaging of diffusion of water into poly(tetrahydrofurfuryl methacrylate-Co-hydroxyethyl methacrylate)", *Polym. Commun.*, vol. 38, no. 15, pp. 3985-3989, 1997. [http://dx.doi.org/10.1016/S0032-3861(96)01071-3]

[40] R. Knauss, G. Fleischer, W. Gründer, J. Kärger, and A. Werner, "Pulsed field gradient NMR and nuclear magnetic relaxation studies of water mobility in hydrated collagen II", *Magn. Reson. Med.*, vol. 36, no. 2, pp. 241-248, 1996. [http://dx.doi.org/10.1002/mrm.1910360211] [PMID: 8843378]

[41] J.E. Tanner, and E. Stejskal, "Restricted self-diffusion of protons in colloidal systems by the pulsed-gradient, spin-echo method", *J. Chem. Phys.*, vol. 49, no. 4, pp. 1768-1777, 1968. [http://dx.doi.org/10.1063/1.1670306]

[42] A. Traore, L. Foucat, and J.P. Renou, " ^{1}H-NMR study of water dynamics in hydrated collagen: transverse relaxation-time and diffusion analysis", *Biopolymers,* vol. 53, no. 6, pp. 476-483, 2000. [http://dx.doi.org/10.1002/(SICI)1097-0282(200005)53:6<476::AID-BIP4>3.0.CO;2-8] [PMID:10775063]

[43] P.T. Callaghan, *Principles of Nuclear Magnetic Resonance Microscopy.* Clarendon Press: Oxford, 2001.

[44] V.V. Rodin, *Magnetic Resonance Methods.* MIPT Press: Moscow, 2004.

[45] V.V. Rodin, P.J. McDonald, and M. Jones, "Two-dimensional distribution function of diffusion in wood obtained using 2D Laplace inversion", *Appl. Phys. Math.,* no. 6, pp. 3-7, 2014.

[46] V.V. Rodin, P.J. McDonald, and S. Zamani, "A nuclear magnetic resonance pulsed field gradient study of self-diffusion of water in hydrated cement pastes", *Diffus. Fundam.,* vol. 18, no. 3, pp. 1-7, 2013.

[47] R.L. Cooper, D.B. Chang, A.C. Young, C.J. Martin, and D. Ancker-Johnson, "Restricted diffusion in biophysical systems. Experiment", *Biophys. J.,* vol. 14, no. 3, pp. 161-177, 1974. [http://dx.doi.org/10.1016/S0006-3495(74)85904-7] [PMID: 4823458]

[48] R. Trampel, J. Schiller, L. Naji, F. Stallmach, J. Kärger, and K. Arnold, "Self-diffusion of polymers in cartilage as studied by pulsed field gradient NMR", *Biophys. Chem.,* vol. 97, no. 2-3, pp. 251-260, 2002. [http://dx.doi.org/10.1016/S0301-4622(02)00078-9] [PMID: 12050014]

[49] J. Hayashi, K. Norinaga, N. Kudo, and T. Chiba, "Estimation of size and shape of pores in moist coal utilizing sorbed water as a molecular probe", *Energy Fuels,* vol. 15, no. 4, pp. 903-909, 2001. [http://dx.doi.org/10.1021/ef0002849]

[50] L. Pel, K. Hazrati, K. Kopinga, and J. Marchand, "Water absorption in mortar determined by NMR", *Magn. Reson. Imaging,* vol. 16, no. 5-6, pp. 525-528, 1998. [http://dx.doi.org/10.1016/S0730-725X(98)00061-7] [PMID: 9803902]

CHAPTER 9

Magnetic Resonance Imaging in Characterisation of Polymer Films

Abstract: Diffusivity and distribution of water in random copolymer films of butyl methacrylate and methacrylic acid swollen in water have been investigated using magnetic resonance imaging (MRI) and NMR relaxation techniques. The relative contributions of the protons of the polymer matrix and bound water have been analyzed using ^{1}H NMR spectra and MRI data of the swollen films. Pulse Field Gradient (PFG) NMR experiments have been carried out in parallel to the MRI studies. The results have been discussed with reference to published MRI data for water in a number of polymer materials. The free induction decay (FID) and the spin-spin relaxation times (T_2) data for water saturated polymer films correlated well with the MRI intensities for the same samples. The results show that by varying the hydrophobic/hydrophilic comonomer ratios in the polymers, the water ingress can be changed. MRI data show how to monitor drying process. Moisture profiles and relaxation data were obtained during film drying. The time dependences of the image slices intensity for the drying films were fitted by a single-exponential function with an average time constant as k = −0.02 min^{-1}.

Keywords: Copolymer swollen films, Evaporation, Free induction decay (FID), Fourier transform (FT), Hydrophobic/hydrophilic ratio, Magnetic resonance imaging (MRI), NMR spectra, Pulsed Field Gradient (PFG) NMR, Spin-spin relaxation time T_2, Stimulated echo (STE), Water.

INTRODUCTION

The polymer films formed from solutions of copolymers are very important in many technologies [1 - 4]. Understanding the ingress of water into copolymer formulations and water effects on the polymer properties are also essential [2, 4 - 12].

Chapter 8 considered the studies of the water state in films formed from the BMA-MAA random copolymer solutions with different mol% of BMA. That research showed how ^{1}H NMR relaxation and PFG NMR techniques could be used to obtain the information about water motion, molecular mobility and interactions [9 - 14]. Chapter 8 also discussed the studies of water self-diffusion in fully dried films that have been swollen in H_2O [2, 4].

Victor V. Rodin

Magnetic resonance imaging (MRI) has a number of significant potential applications in the non-invasive monitoring of absorbed fluids in solid polymers [15 - 17] and in porous materials [8, 9, 13, 15, 18]. MRI studies have been used to investigate the diffusion of water into copolymers with hydroxyethyl methacrylate (HEMA) [19 - 22]. The mass diffusion coefficient of water in polyHEMA at 37°C could be determined from the profiles of the diffusion front [19 - 21].

The present chapter focuses on the MRI applications in studying interactions of water with films formed from the butylmethacrylate (BMA)-polymethacrylyc acid (PMAA) random copolymer solutions with different mol% of BMA. MRI and NMR relaxation techniques are used to explore the water swollen structure and the distribution of water in the films. The free induction decay (FID) and the spin-spin relaxation times (T_2) data in water saturated polymer films correlated well with MRI intensity at varying hydrophobic/hydrophilic comonomer ratios in the polymers. Both sorption and drying processes are described using the MRI data.

POLYMER FILMS SWOLLEN IN WATER AS STUDIED BY FIDS, T_2 AND MRI

The films were prepared from solutions of BMA-MAA random copolymers in IPA as it was described in Chapter 8 and Ref. [4]. The photos of some films under study (with and without water) are presented in Fig. (**1**), Chapter 8. All images were obtained using a Bruker Avance 400 spectrometer operating at 400 MHz for protons and equipped with a micro-imaging accessory and a vertical bore superconductivity magnet. A 5 mm probe was used to image the sample. The sizes of the film samples were 1×5×10 mm. The samples were extracted from the water and placed in NMR tubes, and then inside the probe in such a way to be confined to the area of the RF coil. The relevant *Bruker* microimaging software was used. To study the evaporation process of water from inside the samples, the films with water were periodically kept in a thermostat oven for fixed times after which these were placed in NMR tubes for the MRI studies. Thus, after the first NMR image had been obtained (the film swollen in water), the sample was dried inside a thermostat oven at 323 K for periods of times of 2, 15, 35 and 67 min (total drying time was equal to 119 min). Then, the sample was left at room temperature for 720 min. The total drying time was 839 min. Images have been recorded after each drying period. The intensity of each pixel of an image in MRI studies defined approximately by $M=M_0\times\exp(-TE/T_2)$ where TE (the echo time) represents the delay between the NMR excitation and the collection of the MRI data. An echo time was a constant value in MRI experiment. Then, the relative intensity of each pixel was characterised by the liquid proton density (M_0) and the local spin-spin relaxation time T_2. Spin-spin relaxation arises from the loss of phase coherence between neighbouring spins. Proton T_2 values are very short (about tens of microseconds) for a solid polymer film without water. There was no image from the dried polymer film. The Multi-Slice Multi-Echo (MSME) image pulse sequence [23] was used with the following acquisition parameters: number of slices = 4; field of view (FOV) = 0.95 × 0.95 mm^2; slice thickness (ST) = 2.76 mm; inter-slice thickness (IT) = 2.91 mm; echo time (TE) = 7.0 ms; recycle delay (RD) = 3.0 s. Each image required about 2 hrs data acquisition.

A typical FID for the copolymer films swollen in water consisted of two clearly discernible parts: a fast decaying component (solid-like) and slower decaying component (liquid like) (Fig. **1**). These could be decomposed into separate components using relaxation equations from Chapters 1, 5, 8 or Refs. [4, 7]. The total time-domain signal is the sum of the signals from both parts. The slow relaxing component is modelled by an exponential function (Chapters 1, 2). In the case of swollen films, a double exponential fit was used to model slow FID component. The analysis of the FIDs gave the initial intensities of the different components which could also be done by integration of the frequency-domain peaks to fit the experimental data. When the receiver dead time is less $3T^*_2$ for the solid, it is possible to isolate the signal from the liquid. The amount of pore liquid could be estimated then on a relative scale. As an additional check, the system was also analyzed by a gravimetric method.

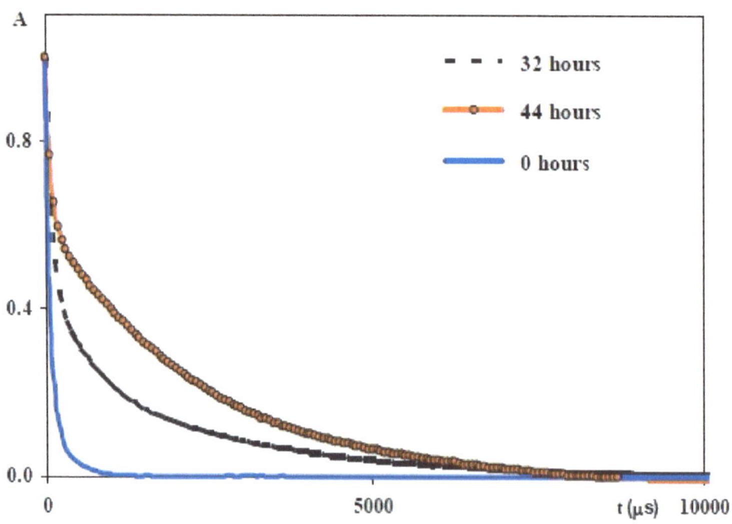

Fig. (1). Typical FIDs in dry BMA-MAA copolymer 60/40 film, and the same film swollen in water for 32 and 44 hours. Solid and dotted lines for the swollen films are the fits with 2 exponential functions.

Water in different physical environments can be characterized by different transverse relaxation time T_2. The CPMG experiment, $90°_x - (t_{cp} - 180°_y - t_{cp})_n$ [7, 14, 23], has been used to quantify the relative proportions of water with different values of T_2. T_2 is shortened from the bulk value by interaction with the walls of the pore film, and T_2 in many porous systems accounted for the pore sizes [4, 13, 24].

MRI data obtained show a "visible" distribution of water within the film (Fig. **2**). The water distribution is as follows: white areas represent regions without water and dark colour spots represent the regions of greatest water concentration in the samples. The images clearly show a very non-uniform distribution of absorbed water which it is seen to be preferentially positioned through the surface and inner layers close to the surface.

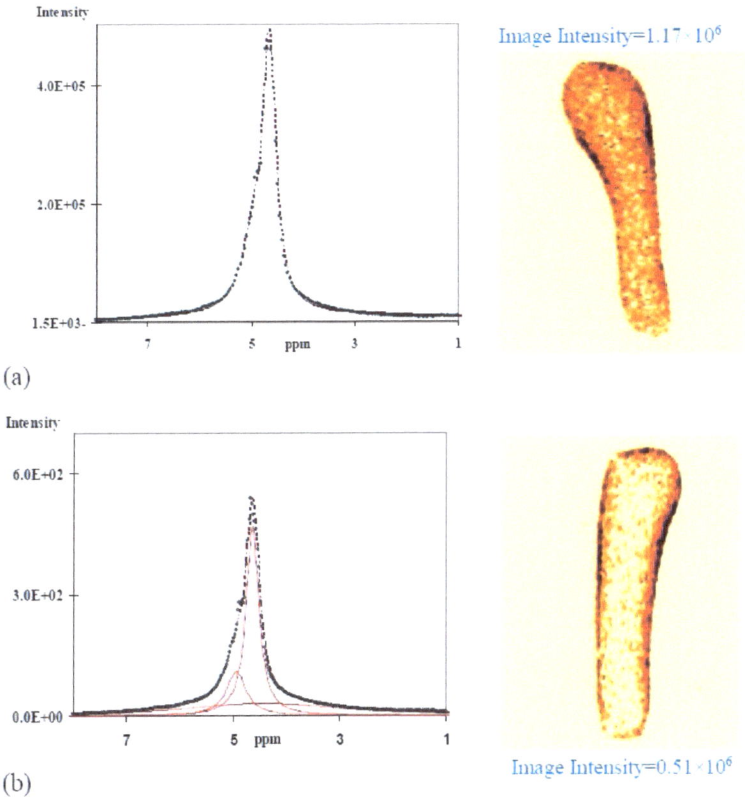

Fig. (2). NMR images and ^{1}H NMR spectra for two BMA-MAA 90/10 films: one film (a) is after 55 days, and other one (b) is after 48 days of water exposure. Dashed lines are the fits with a sum of the 3 Lorentz peaks (with line widths ω_{L1}, ω_{L2} and ω_{L3}). Solid lines (detail fit is shown for a film (b)) are 3 Lorentz modeled peaks (with line widths ω_{L1}, ω_{L2} and ω_{L3}) used in fitting the experimental ^{1}H NMR spectra.
a: $\omega_{L1}\approx42$ Hz, $\omega_{L2}\approx65$ Hz and $\omega_{L3}\approx600$ Hz.
b: $\omega_{L1}\approx43$ Hz, $\omega_{L2}\approx85$ Hz and $\omega_{L3}\approx1000$ Hz.

Figs. (**2a, b**) show the NMR images data and ^{1}H NMR spectra for the wet films BMA-MAA 90/10 after different water exposure. There are some striking features. The wetter film has an image intensity that exceeds that for the film with shorter water exposure. Thus, these water exposure conditions are far from saturation. Comparison of the ^{1}H NMR spectra for these wet films shows that for the polymer film with a narrow peak in the ^{1}H NMR spectrum it is possible to get detailed image data. If the film is characterized by slightly broadening NMR spectrum, then the intensity of the T_2 image is less. This is why the second example in Fig. (**2**) shows white regions (without water) in the center of the film. A higher water concentration is present in the surface of the films. The total image intensity was larger for a wetter film for the same experimental conditions.

A deconvolution of the NMR spectra of these wet 90/10 BMA-MAA films has been carried out with fitting a sum of three Lorentz peaks (with line widths ω_{L1}, ω_{L2} and ω_{L3}). Narrow peaks (with line widths ω_{L1}, ω_{L2}) characterize mobile water inside the film (their line widths correspond to a long T_2 = 3.7–7.6 ms). These signals account for most of the image intensity.

The relative integral intensity (calculated as a fraction of all integrated intensity of ^{1}H NMR spectra) for a sum of narrow peaks (with lines ω_{L1}, ω_{L2}) correlates with the image intensity of the films. For example, for one 90/10 BMA-MAA film, this fraction (for lines ω_{L1}, ω_{L2}) is equal to 62%, and for the 90/10 BMA-MAA film with a more intensive image, that is about 76% (Fig. **2**). In addition to this, the image intensities of the wet films correlated with the water self-diffusion constants (obtained from PFG NMR). Fig (**3**) presents one set of these data. For water inside the film BMA-MAA 90/10 with intensive image, an apparent diffusion coefficient (D= $0.78 \cdot 10^{-12}$ m^2/s) is higher than diffusion constant (D= $0.42 \cdot 10^{-12}$ m^2/s) in the film BMA-MAA 90/10 with lower intensity image.

It was shown that with increasing the mol% MAA in the BMA-MAA copolymer films, a fraction of mobile protons (after holding the films in water) increased. This was confirmed by the measurements of the FID curves with much slower T_2 components and by the NMR spectra with narrower peaks. To understand the role of the mol% BMA in the copolymer film for water uptake, films with mol% BMA = 60 and mol% BMA=100 have been additionally studied. Fig. (**4**) shows the NMR images for the BMA-MAA 60/40 film and for the film from pure PBMA after water exposure for the same time and their respective ^{1}H NMR spectra. It seems that hydrophobic polymer film (pure PBMA) does not absorb very much water (poor image) under similar experimental conditions to the BMA-MAA 60/40 film as might be expected. It is also possible to observe in the image that the sample has several cracks, and the water is located only in the areas close to the cracks. The presence of the cracks through the center of the film provides a

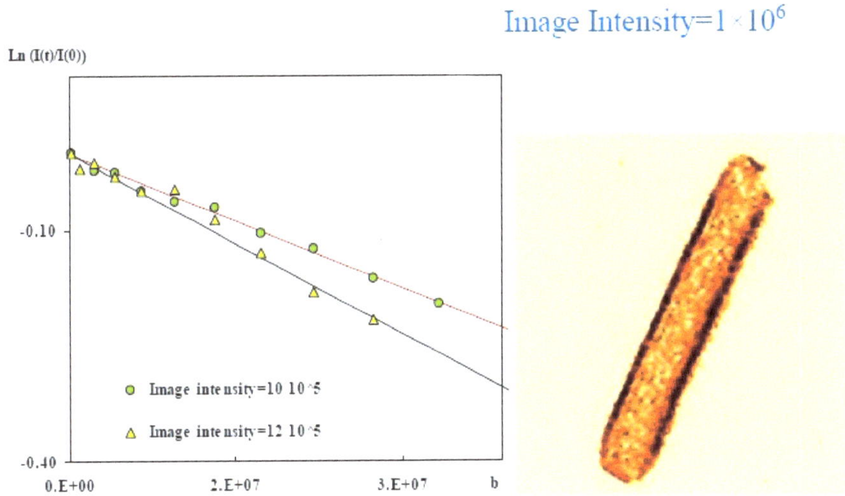

Fig. (3). Typical attenuation plots of echo decays in PFG NMR experiments as a function of gradient factor ***b*** for two wet films BMA-MAA 90/10 with image intensity = 1.2×10^6 (after 55 days of water exposure) and image intensity = 1×10^6 (after 50 days of water exposure). Diffusion time Δ=100 ms. G_{max}=10 T/m. Image (one slice from 8 slices set of MRI experiment) is shown for the film with image intensity =1×10^6.
Solid lines are the fits to a single exponential function giving the diffusion constants $D = 0.78 \cdot 10^{-12}$ m²/s (film with image intensity = $1.2 \cdot 10^6$) and $D = 0.42 \cdot 10^{-12}$ m²/s (film with image intensity = 1×10^6).

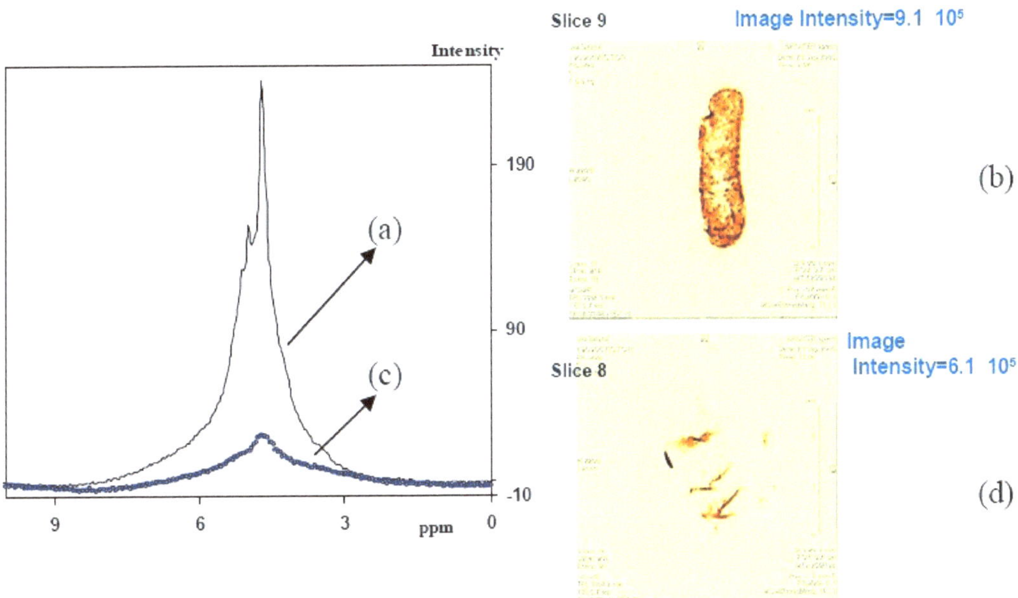

Fig. (4). ^{1}H NMR spectra (a, c) and NMR images (b, d) for two wet films: BMA-MAA 60/40 (a, b) and 100/0 (c, d) after 50 days water exposure.

mechanism for macroscopic transport of the water deeper into the center of the film. The transfer of water through crack channels (*e.g.* scratches) was shown previously for a HEMA copolymer in water by MRI [19, 20]. At the same time, some parts of the pore structure were left without water. These are seen as white regions in the NMR images. Thus, the presence of defects in the polymer matrix, such as cracks, may explain the existence of a mobile phase of absorbed water within the polymer. The existence of cracks within the polymer provides an alternative route for the transfer of water. However, in cases where MRI did not show the cracks, the diffusion of water inside the film can only be explained by ingress into the pore structure of samples.

The experimental FIDs for the BMA-MAA copolymer films swollen in water for longer exposure showed a large portion of mobile protons (Fig. **1**). With increasing the mol% BMA in the BMA-MAA copolymers, this fraction of mobile protons decreased.

In order to obtain more details about the role of the BMA monomers in water uptake, a comparative study of the BMA-MAA 60/40 and pure PBMA films has been carried out [9]. For example, after 6 hrs water exposure, the polymer film (mol% BMA = 60) had 57% of protons with T_2 value characterising mobile phase. However, in a parallel study, the film with pure PBMA after water exposure for 6 hrs had a lower population of protons (30%) with comparable T_2 values. This confirmed that interaction between water and these copolymer films is determined by the hydrophilic/hydrophobic ratios [4, 9, 25]. Published data [19 - 21] showed that the effect of water absorption in the copolymers and plasticization depend on the fraction of hydrophilic domains in the polymer chain. For example, Ghi *et al.* [19, 20] carried out NMR imaging studies on a series of copolymers of hydroxyethyl methacrylate (HEMA) and tetrahydrofurfuryl methacrylate (THFMA) and showed that the rate of diffusion of water into a copolymer, as well as the amount of equilibrium water which can be absorbed by the polymer matrix, was dependent on the molar ratio of the monomers.

The dynamics of water molecules inside the film depends on their chemical surrounding and the structure of the film. Water is a nonsolvent for BMA, but it is a solvent for MAA units [4, 15, 25, 26]. At longer exposure times, faster molecular motion was detected for the water in the film matrix. For example, in the case of 12 hrs water exposure, the water molecules inside the film have correlation times τ_c that are smaller than those estimated inside the film after 6 hrs water exposure. After 12 hours of holding the BMA-MAA copolymer films in water (mol% BMA = 60 and 80), the CPMG curves measured on these films could be fitted by two exponential components. The fast relaxing component was characterised by correlation times of τ_c = (0.42–0.74)×10^{-8} s. A slow relaxing component was responsible for the molecular motions with $\tau_c = (1.2–1.6) \times 10^{-10}$ s. The last values of τ_c for mobile water inside the polymer film can be compared with published values of τ_c (~ 1×10^{-10} s) for water adsorbed in natural polymers at a similar hydration level (HL ~ 0.2 g H_2O per g dry matter) [7, 27]. This slow relaxing water component is consistent with water in the pores of the microstructure of the polymer (natural or synthetic) network.

NMR AND MRI IN STUDYING EVAPORATION OF WATER FROM THE FILMS

Fig. (**5**) shows ^{1}H NMR spectra for one particular case of BMA-MAA 90/10 swollen film whilst drying at T = 318 K. It shows that the signal intensity of the mobile water decreases with increasing drying time.

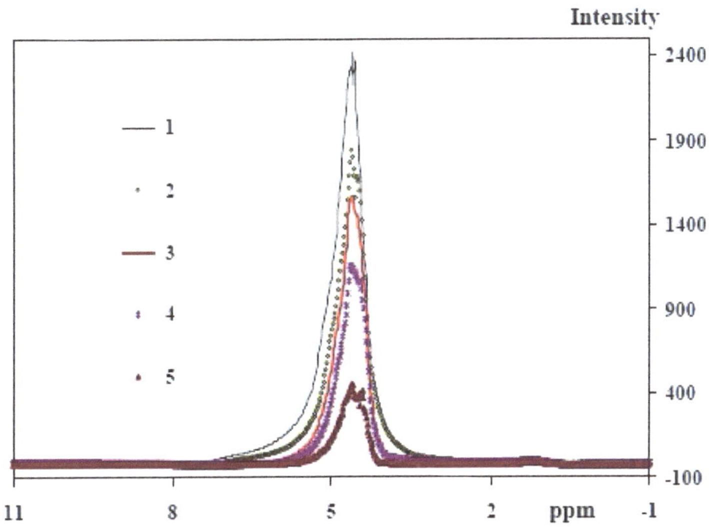

Fig. (5). ^{1}H NMR spectra for the BMA-MAA 90/10 swollen film during drying at T = 318 K. Drying time is increasing from step 1 to step 5 [25]: (1) – 0 min ; (2) – 18 min; (3) – 228 min; (4) – 318 min; (5) – 548 min;.

During the time of this drying, the water content of the BMA-MAA 90/10 film (weight of water per weight of sample, H_w, in %) decreased from ~20% to 2.5%.

The decay of spin-echo signal with time measured in CPMG experiment on this BMA-MAA 90/10 film was fitted by two relaxation components [25]. Fig. (**6**) shows how the spin-spin relaxation times (long T_2' and short T_2'') of these components and their intensities changed with drying this BMA-MAA 90/10 film. The water component with the long T_2 value (associated with water molecules in large pores) decreased by a factor of 3 after 220 min drying. The component with the smaller T_2 value (associated with water in the smaller pores) decreased by a factor of 2 times after 220 min drying. This suggests that water molecules from the large pores leave the swollen film faster than water molecules from smaller pores during drying. The final ^1H spectra of the films (after drying) and the spectra of original dried films (before swelling in water) are comparable. Thus, the films swollen in water could be returned back to their original dried state.

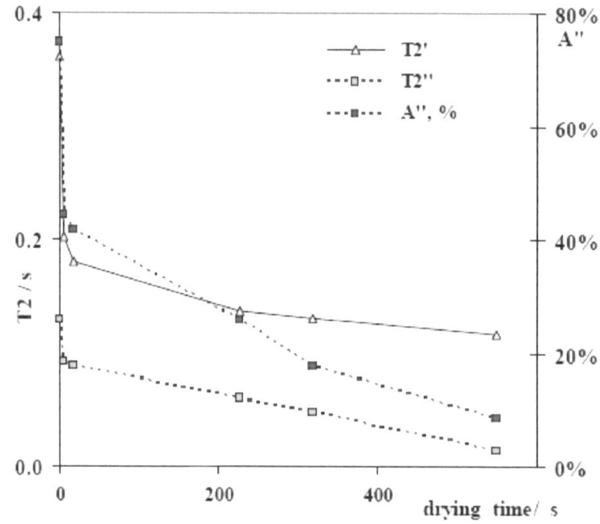

Fig. (6). The spin–spin relaxation times for two components (long – T_2' and short – T_2'') and relative intensity of the fast decaying component (% of total measurable signal) *vs* drying time of the BMA-MAA 90/10 film after water exposure. Drying was at 318 K. Original water content H_w = 19.9 %.

MRI has been applied in recording images of the BMA-MAA 60/40 copolymer film (swollen in water for 6 days) when the sample was progressively dried at 323 K in a convection oven. At various drying steps, the sample was removed from the oven, placed into NMR tube, and measured for this water content using MSME pulse sequence. MSME images obtained in this study were recorded for 4 slices with a thickness of 2.76 mm. Fig. (**7**) presents the images of first slices for several particular steps of progressive drying the BMA-MAA 60/40 film. The signal required to produce the images comes from areas that contain water only. The detectable areas in these images get smaller as the sample is getting drier. Fig. (**7**) shows also that the image at time = 0 has a larger intensity at the edges than that in the centre. This is due to the fact that the water has not saturated the sample after 6 days of immersion, and diffusion is still an ongoing process at this time. As the sample is drying, the signal intensity becomes more uniform over the sample. The images have been integrated, and all integral intensities were analyzed for all stages of evaporation. Integrated image intensities on four slices of each experiment with particular drying time were averaged and then have been considered as a function of the drying time (evaporation). The image intensity is damping as follows: $I \approx I_0 \times \exp(-0.02 \times t)$ where drying time [t] is expressed in minutes.

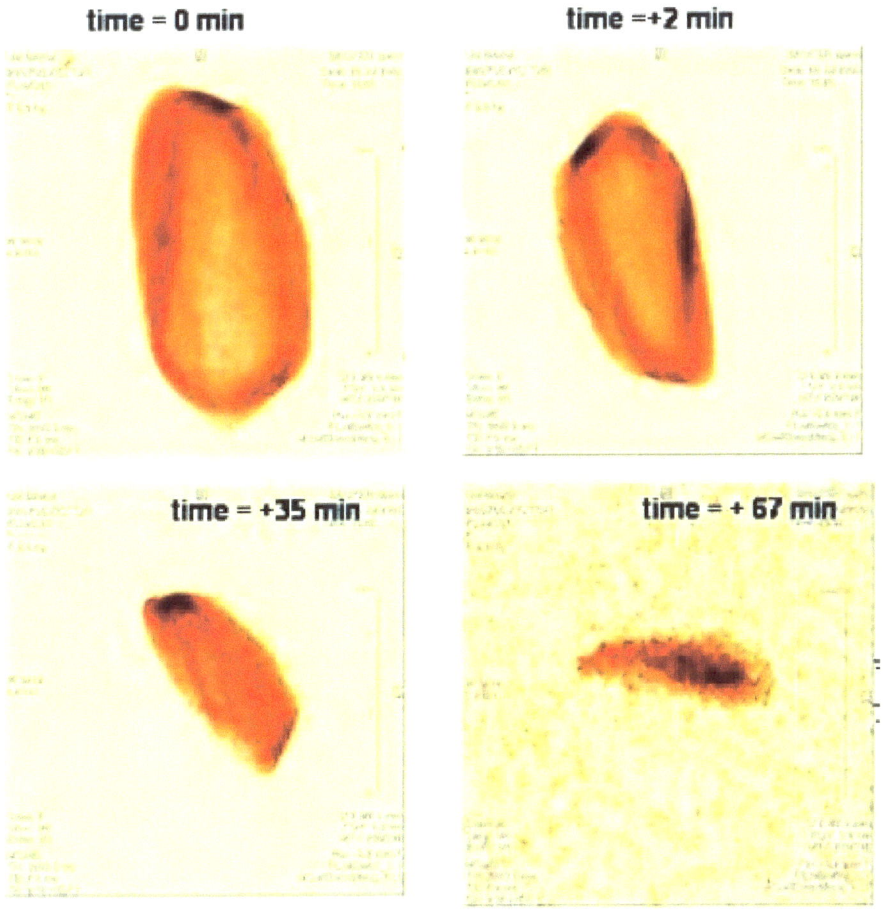

Fig. (7). NMR images recorded for BMA-MAA 60/40 copolymer film swollen in water for 6 days. 4 examples of progressively drying the film are shown. The images are presented for 1st slice in each data set of 4 slices. MRI measurements have been done at room temperature, and the frequency is 400 MHz.

The integrated intensities for the four slices of each experiment as a function of the drying time have been also analyzed separately. The fits for each data set were carried out to an exponential function. Three of the fittings gave the same value for the rate of decay, but the one corresponding to the top slice gave a slightly slower decay, with values of −0.021 and −0.017 min^{-1}, respectively. After more than 300 min drying the film swollen in water, it could be returned back to the original dried state (before water exposure). Thus, it is possible to eliminate water from the film by keeping it opened to air or putting it in an oven at T= 323 K. The

^{1}H spectra of such films and the spectra of original dried films (before swelling in water) were similar.

CONCLUDING REMARKS

NMR Imaging and the NMR relaxation techniques have been used to characterize water in butyl methacrylate and methacrylic acid copolymer films. FID, T_2 and MRI experiments gave water concentration distributions at various molar BMA-MAA ratios and under different immersion conditions. After increasing the exposure time of the films in water, an additional increase in water absorption and mobility of water molecules in the films was observed. At an increase of hydrophilic monomer MAA fraction in the copolymer films, the water uptake became more intensive at the same water exposure time.

Diffusion coefficients of water in swollen BMA-MAA films correlated with the MRI intensities of swollen film. The transverse relaxation gave information on the distribution of water in the inner pore structure. ^{1}H spectra confirmed that the films swollen in water could be returned back to their original dried state. The volume averaged magnetization decays, ^{1}H NMR spectra as well as the T_2 relaxation times in polymer films swollen in water correlated with the intensities of the NMR images for the same samples. The results demonstrated that modification of BMA-MAA copolymer films by changing the hydrophobic/hydrophilic monomers influences on the water uptake in the porous polymer materials.

ACKNOWLEDGEMENTS

The author thanks Terence Cosgrove and Martin Murrey.

REFERENCES

[1] P. Broz, *Polymer-Based Nanostructures: Medical Applications.* Royal Society of Chemistry: London, UK, 2010.
[http://dx.doi.org/10.1039/9781847559968]

[2] W.J. Soer, J. Scheerder, G. Satgurunathan, Y. Liu, V.V. Rodin, and J.L. Keddie, "Water transport through polymer films: comparison of solution, emulsion and secondary emulsion polymers", *International Conference (20-24 July 2014, Prague) 'Frontiers of Polymer Colloids: From Synthesis to Macro-Scale and Nano-Scale Applications',* 2014, p. 191

[3] H. Yazdani-Ahmadabadi, S. Rastegar, and Z. Ranjbar, "Slip-Stick Diffusion Behavior Caused by Photo-Polymerization-Induced Nano-Gelation in Highly Heterogeneous Photo-Polymeric Coatings", *Prog. Org. Coat.,* vol. 97, pp. 184-193, 2016.
[http://dx.doi.org/10.1016/j.porgcoat.2016.04.010]

[4] V.V. Rodin, and T. Cosgrove, "Nuclear magnetic resonance study of water-polymer interactions and self-diffusion of water in polymer films", *OALJ: Chem. Mater. Sci,* vol. 3, no. 10, pp. 1-17, 2016.

[5] P. Ducheyne, K.E. Healy, D.E. Hutmacher, D.W. Grainger, and C.J. Kirkpatrick, *Comprehensive Biomaterials* vol. 1. Elsevier Ltd, 2011, p. 3672.

[6] Y. Liu, A. Gajewicz, V.V. Rodin, W-J. Soer, J. Scheerder, G. Satgurunathan, P.J. McDonald, and J.L. Keddie, "Explanations for water whitening in secondary dispersion and emulsion polymer films", *J. Polym. Sci., B, Polym. Phys.,* vol. 54, no. 16, pp. 1658-1674, 2016. [http://dx.doi.org/10.1002/polb.24070]

[7] V.V. Rodin, "Methods of Magnetic Resonance in Studying Natural Biomaterials", In: *Encyclopedia of Physical Organic Chemistry,* Zerong Wang, Ed., Chapter 53. vol. 4, part 4. John Wiley & Sons, Inc., 2017, pp. 2861-2908.

[8] K.N. Allahar, B.R. Hinderliter, D.E. Tallman, and G.P. Bierwagen, "Water transport in multilayer organic coatings", *J. Electrochem. Soc.,* vol. 155, no. 8, pp. F201-F208, 2008. [http://dx.doi.org/10.1149/1.2946429]

[9] V.V. Rodin, T. Cosgrove, M. Murray, and R. Buscall, "Intermolecular interactions upon film formation from copolymer solutions", *Book of Abstracts XVIIth European Chemistry at Interfaces Conference,* 2005, p. 107, Loughborough.

[10] J. Rottstegge, K. Landfester, M. Wilhelm, H.W. Spiess, and C. Heldmann, "Different types of water in the film formation process of latex dispersions as detected by solid-state nuclear magnetic resonance spectroscopy", *Colloid Polym. Sci.,* vol. 278, no. 3, pp. 236-244, 2000. [http://dx.doi.org/10.1007/s003960050037]

[11] V. Baukh, H.P. Huinink, O.C. Adan, S.J. Erich, and L.G. van der Ven, "Predicting water transport in multilayer coatings", *Polymer,* vol. 53, no. 15, pp. 3304-3312, 2012. [http://dx.doi.org/10.1016/j.polymer.2012.05.043]

[12] V. Baukh, H.P. Huinink, O.C. Adan, S.J. Erich, and L.G. van der Ven, "Water-polymer interaction during water uptake", *Macromolecules,* vol. 44, no. 12, pp. 4863-4871, 2011. [http://dx.doi.org/10.1021/ma102889u]

[13] P.T. Callaghan, *Principles of Nuclear Magnetic Resonance Microscopy.* Clarendon Press: Oxford, 2001.

[14] P.T. Callaghan, *Translational Dynamics and Magnetic Resonance.* Oxford University Press: Oxford, 2011. [http://dx.doi.org/10.1093/acprof:oso/9780199556984.001.0001]

[15] W.P. Rothwell, D.R. Holecek, and J.A. Kershaw, "NMR imaging: study of fluid absorption by polymer composites", *J. Polym. Sci. Polym. Lett. Ed.,* vol. 22, no. 5, pp. 241-247, 1984. [http://dx.doi.org/10.1002/pol.1984.130220501]

[16] G. Laplante, A.E. Marble, B. Macmillan, and P.L. Sullivan, "Detection of water ingress in sandwich structures: a magnetic resonance approach", *NDT Int.,* vol. 38, no. 6, pp. 501-507, 2005. [http://dx.doi.org/10.1016/j.ndteint.2005.01.006]

[17] A.E. Marble, G. Laplante, I.V. Mastikhin, and B.J. Balcom, "Magnetic resonance detection of water in composite sandwich structures", *NDT & E. Int.,* vol. 42, no. 5, pp. 404-409, 2009. [http://dx.doi.org/10.1016/j.ndteint.2009.01.010]

[18] A.P. Lehoux, S. Rodts, P. Faure, E. Michel, D. Courtier-Murias, and P. Coussot, "Magnetic resonance imaging measurements evidence weak dispersion in homogeneous porous media", *Phys Rev E,* vol. 94, no. 5, p. 053107, 2016. [http://dx.doi.org/10.1103/PhysRevE.94.053107] [PMID: 27967061]

[19] P.Y. Ghi, D.J. Hill, and A.K. Whittaker, "NMR imaging of water sorption into poly(hydroxyethyl methacrylate-co-tetrahydrofurfuryl methacrylate)", *Biomacromolecules,* vol. 2, no. 2, pp. 504-510, 2001. [http://dx.doi.org/10.1021/bm000146q] [PMID: 11749213]

[20] P.Y. Ghi, D.J. Hill, D. Maillet, and A.K. Whittaker, "NMR imaging of diffusion of water into poly(tetrahydrofurfuryl methacrylate-Co-hydroxyethyl methacrylate)", *Polym. Commun.,* vol. 38, no. 15, pp. 3985-3989, 1997.

[http://dx.doi.org/10.1016/S0032-3861(96)01071-3]

[21] M.A. Chowdhury, D.J. Hill, and A.K. Whittaker, "NMR imaging of the diffusion of water at 310 K into poly[(2-hydroxyethyl methacrylate)-co-(tetrahydrofurfuryl methacrylate)] containing vitamin B-12 or aspirin", *Polym. Int.,* vol. 54, no. 2, pp. 267-273, 2005. [http://dx.doi.org/10.1002/pi.1665]

[22] M.A. Chowdhury, D.J. Hill, and A.K. Whittaker, "NMR imaging of the diffusion of water at 37 ° C into Poly(2-hydroxyethyl methacrylate) containing aspirin or vitamin B(12)", *Biomacromolecules,* vol. 5, no. 3, pp. 971-976, 2004. [http://dx.doi.org/10.1021/bm030079a] [PMID: 15132689]

[23] J. Tritt-Goc, N. Piślewski, A. Kaflak-Hachulska, D. Chmielewski, A. Górecki, and W. Kolodziejski, "Proton magnetic resonance microimaging of human trabecular bone", *Solid State Nucl. Magn. Reson.,* vol. 15, no. 2, pp. 91-98, 1999. [http://dx.doi.org/10.1016/S0926-2040(99)00032-6] [PMID: 10670900]

[24] L. Pel, K. Hazrati, K. Kopinga, and J. Marchand, "Water absorption in mortar determined by NMR", *Magn. Reson. Imaging,* vol. 16, no. 5-6, pp. 525-528, 1998. [http://dx.doi.org/10.1016/S0730-725X(98)00061-7] [PMID: 9803902]

[25] V.V. Rodin, T. Cosgrove, and M. Murrey, "Magnetic resonance imaging and pulsed field gradient NMR in study of structure of polymer solutions and films", *Proceedings of the 6th Colloquium on Mobile Magnetic Resonance,* 2006, pp. 32-33 Aachen, Germany.

[26] T. Cosgrove, V.V. Rodin, M. Murray, and R. Buscall, "Self-diffusion in solutions of carboxylated acrylic polymers as studied by pulsed field gradient NMR. 2. Diffusion of macromolecules", *J. Polym. Res.,* vol. 14, no. 3, pp. 175-180, 2007. [http://dx.doi.org/10.1007/s10965-006-9088-0]

[27] M-C. Vackier, B.P. Hills, and D.N. Rutledge, "An NMR relaxation study of the state of water in gelatin gels", *J. Magn. Reson.,* vol. 138, no. 1, pp. 36-42, 1999. [http://dx.doi.org/10.1006/jmre.1999.1730] [PMID: 10329223]

CHAPTER 10

Self-Diffusion of Water in Cement Pastes as Studied by ^{1}H PFG NMR and DDCOSY NMR

Abstract: The chapter considers the data of one-dimensional ^{1}H NMR diffusion studies in white Portland cement paste (the water-cement ratio (w/c) = 0.4) with the age of 1 to 14 days and 1 year. Two-dimensional diffusion-diffusion correlation (DDCOSY) experiments have been carried out in cement paste (w/c= 0.4) aged from 1 to 7 days. It was found that the ^{1}H pulsed field gradient (PFG) NMR experiments measure diffusivities of capillary water. A log-normal pore size distribution and relaxation times dependent on pore size have been applied to fit the data. Mature paste has been characterized by mean capillary pore size of 4.2 μm. The data for cement pastes with the age of 1 week showed similarity. On the basis of data obtained it was suggested that gel porosity and hydrates do not form in the capillary porosity. 2D DDCOSY experiments did not discover any diffusion anisotropy in capillary pores.

Keywords: 2D diffusion-diffusion correlation NMR spectroscopy (DDCOSY), Apparent diffusion coefficient D_{app}, Cement paste, Diffusion time (Δ), Inverse Laplace transform (ILT), Length of gradient pulse (δ), Pulsed Field Gradient (PFG) NMR, Restricted diffusion, Spin-lattice relaxation time T_1, Spin-spin relaxation time T_2, Stimulated echo (STE), Water.

INTRODUCTION

Hardening cement paste is a very complex system consisting of nonhydrated cement and hydration products [1 - 8]. A detailed understanding of the evolution of pore microstructure in cements, and of water dynamics at different scales within cement pores remains very poorly understood [1 - 4, 7 - 10]. Cement pastes contain many crystalline phases. The microstructure of hydrating cement paste determines largely the technical properties of concrete, *e.g.* permeability, durability and strength. The formation of gel with calcium-silicate-hydrate (C-S-H) matrix is the principal factor in the setting and hardening of the cement paste [1, 4, 8, 10 - 12]. C-S-H gel is the most significant hydration product of cementitious materials. C-S-H is responsible for the properties of the fully hydration cement paste [4, 8, 10].

Victor V. Rodin

Different methods (including NMR) have been performed in cement studies [1 - 5, 12]. For example, 2D NMR T_2–T_2 relaxation exchange experiments [8 and Refs. there in] in white Portland cement pastes have confirmed a hierarchy of pore sizes and directly measured the exchange rate for water between them. On multiple length scales, cement is characterised as heterogeneous and complex material. PFG NMR is used for studying self-diffusion of molecules in solutions or complex systems [2, 12, 13]. An apparent diffusion coefficient D_{app} for liquids confined in porous media can depend on diffusion time and can inform about diffusion with restrictions. This can detail the microstructure and estimate the sizes of confinement [13 - 15]. Some kinds of cements have been already studied by PFG NMR techniques [2, 11, 12]. Nevertheless, many issues of NMR diffusion applications in cements are still in challenges. For example, quantitative measurements can be difficult if the applied field gradients overlap with large magnetic field gradients due to the inhomogeneity of the magnetic susceptibility of the sample. In addition to this, there are small pores in the cement paste [1, 4 - 9]. The times of spin-spin and spin-lattice relaxation are very short in cements. This limits the maximum length of gradient pulse and the diffusion time. Therefore, first of all, self-diffusion of water in large, *i.e.*, capillary pores can be studied by this method.

The chapter presents the results of NMR diffusion studies (one-dimensional PFG and 2D diffusion-diffusion correlation spectroscopy) in white Portland cement pastes which were cured for 1 to 14 days and 1 year. The known equation describing a restricted diffusion in the case of a Gaussian pore size distribution [16], has been adapted to the logarithmic-normal type of pore size distribution with spin-lattice and spin-spin relaxation times depended on the pore size. The sizes obtained for the log-normal distribution had micron size that characterized the capillary pores. According to suggestions [17, 18], the spaces of C-S-H gel are mostly planar. Two-dimensional PFG NMR spectroscopy has been used looking for any anisotropic motion in capillary water and in the nascent gel of cement pastes at early stages of hydration.

METHODS AND MODELS

The working frequency for protons in all NMR experiments was 400 MHz. The methods applied in 1- and 2-dimensional studies of cement pastes were considered in Chapter 1, Fig. (**7**) and in [2]. The basis of these methods is an application of two gradient pulses to the radio frequency pulse sequence of spin-echo (SE) or stimulated echo (STE): G is a strength of gradient pulses, δ is a duration of gradient pulses, and Δ is a time separation between the first edges of gradient pulses.

Methods of Spin-Echo and Stimulated Echo in PFG Experiments

In one-dimension SE pulse sequence was used with echo time ranging from 12 to 36 ms, gradient pulses of duration 2 ms, and maximum amplitude G_m=1.5 T/m. Typically 512 to 1024 averages with repetition time of 1 s were recorded per each echo spectrum. T_2 distributions in cement pastes were measured using CPMG experiments with the pulse spacing of 25 μs and 200 echoes. More details are presented in Chapter 1 and Refs. [2, 8].

When two field gradients are used in SE and STE pulse sequences, the intensity of measured echo becomes smaller in comparison with the echo in these experiments without gradients. This attenuation is due to diffusion of molecules. For the case of small molecules in liquids, the diffusion attenuation of echo signal is given in Chapter 1 by eq. (6). Tanner studied both SE and STE pulse sequences for many practical cases (in particular, whenever $T_1>T_2$) and analyzed the attenuation of the spin-echo in the presence of inherent gradient [23 - 25]. The results showed that if the presence of an inherent gradient g_0 cannot be avoided, the attenuation of the echo is given for the stimulated echo pulse sequence, respectively, by eq. (1) [24]:

$$\ln\left(\frac{M}{M_0}\right)=-\gamma^2 D\left\{\delta^2\left(\Delta-\frac{1}{3}\delta\right)G^2+\tau_{1t}^2(\tau_{2t}-\frac{1}{3}\tau_{1t})g_0^2-\delta\left[t_1^2+t_2^2+\delta(t_1+t_2)+\frac{2}{3}\delta^2-2\tau_{1t}\tau_{2t}\right]Gg_0\right\} \quad \textbf{(1)}$$

Here, γ is the gyromagnetic ratio for protons; D is the diffusion coefficient; M_0 is the initial nuclear signal and M is that at the time of echo maximum. τ_{1t} is the time when 2nd (π/2) RF pulse occurs, whereas τ_{2t} is the time of switching 3rd (π/2) RF pulse. The first gradient pulse occurs at a time t_1. The time between the end of the 2nd gradient pulse and the maximum of the spin-echo is t_2. In the case when $\tau_{1t}=\tau_{2t}=\tau$ (*i.e.*, STE pulse sequence becomes as SE diffusion experiment indeed), eq. (1) results in final expression for diffusion attenuation of the normal echo, which is identical to the equation previously presented by Tanner for SE diffusion experiment [25]:

$$\ln\left(\frac{A(2\tau)}{A_0}\right)=-\gamma^2 D\left\{\delta^2\left(\Delta-\frac{1}{3}\delta\right)G^2+\frac{2}{3}\tau^3 g_0^2-\delta\left[t_1^2+t_2^2+\delta(t_1+t_2)+\frac{2}{3}\delta^2-2\tau^2\right]Gg_0\right\} \quad \textbf{(2)}$$

For SE diffusion pulse sequence, the first gradient pulse occurs at a time t_1 and the second gradient is switched on at a time $\Delta+t_1$. For each gradient $G(t)$, spin-echo appears at $t=2\tau$ giving the effect on the intensity of echo signal $A(2\tau)$ according to eq. (2). A_0 is the amplitude of spin-echo in absence of gradients. If there is no applied gradient G, only the term in g_0^2 remains. Then, the result is the same as it is obtained for normal two pulse spin-echo experiment [14, 21]. When g_0 is close to zero or even vanishes, only the term in G^2 remains and final expression results in eq. (3):

$$\ln\left(\frac{A(2\tau)}{A_0}\right) = -\gamma^2 D\delta^2\left(\Delta - \frac{1}{3}\delta\right)G^2 \qquad \textbf{(3)}$$

This result is similar to that presented in Chapter 1 for both STE and SE diffusion experiments in absence of background gradients.

In T_2 experiment, the background gradients can result in a decrease in the measured spin-spin relaxation times. The background gradients may complicate the use of NMR diffusion measurements in some materials. To reduce the effect of inherent gradients in diffusion studies, it was suggested to apply the alternating magnetic field gradients [26]. The authors of this work have shown that at applying the bipolar (alternating) field gradients, the contribution from the cross term Gg_0 between the applied gradient G and the inherent gradient g_0 can be eliminated. That pulse sequence [26] used pulsed field gradients of both polarities instead two applied gradient pulses of the same polarity with one in the prepare gap and the other in the read interval. Cotts *et al.* [19] described three new STE pulse sequences (nine-, 13- and 17-intervals) showing that the echo-attenuation and systematic error caused by both terms (Gg_0 and g_0^2) are significantly reduced. The additional pulses (for example, in 13-interval pulsed sequence in comparison with nine-interval diffusion experiment with bipolar gradients) greatly increase the echo-attenuation caused by the G^2 term, whereas the cross term may vanish at correct choice of conditions.

If the 13-interval *Cotts* pulsed sequence is applied in NMR diffusion studies, the echo-attenuation can be presented as eq. (5) in [19] or eq. (2) in [20]. The equation contains a term with the square of the background gradient, a cross term Gg_0, and a term with the square of the applied gradient. Last term is similar to eq. (3) (above) or to eq. (6) in Chapter 1. The notation in this equation, and in Ref. [20] follows to that which was adopted in Callaghan's and Tanner's publications [21 - 23]. However, the authors of Ref. [19] used Δ to denote the time interval between 2nd (π/2) RF and 3rd (π/2) RF pulses. The echo-attenuation measured in the 13-interval pulsed sequence can be presented (using Tanner's notation) according to eq. (4):

$$\frac{I(\delta, G, \delta_1, \delta_2, g_0)}{I(0)} = \exp\left\{-\gamma^2 D\left[\delta^2\left(4\Delta - \tau_1 - \frac{2}{3}\delta\right)G^2 + \tau_1\delta(\delta_1 - \delta_2)Gg_0 + \frac{1}{6}\tau_1^3 g_0^2\right]\right\} \qquad \textbf{(4)}$$

It is assumed [19] that background gradient g_0 is constant over the diffusion path of considered nucleous. τ_1 is the time when 2nd ($\pi/2$) RF pulse occurs; τ_2 is the time between the 2nd ($\pi/2$) RF and the 3rd ($\pi/2$) RF pulses. δ_1 is the time interval from the first radiofrequency pulse to the start of the gradient; δ_2 is the time interval following the end of the gradient pulse to the next (π or π /2) RF pulse. The same δ_1 and δ_2 intervals (before and after each gradient pulse) apply to all other gradient pulses also. The 13-interval pulsed sequence with bipolar gradients implemented for NMR diffusion studies of cements is shown in Fig. (**1**).

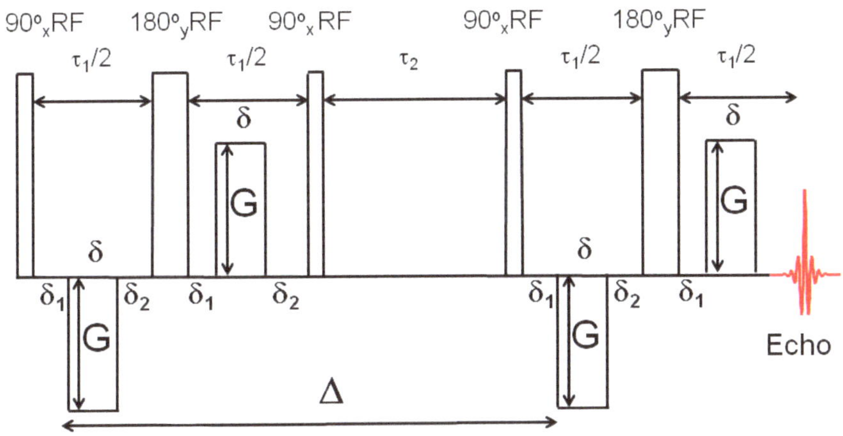

Fig. (1). The thirteen-interval pulse sequence according to [19]. The time intervals (δ_1 and δ_2) before and after bipolar gradient pulse apply to all gradient pulses.

The expressions describing the echo-attenuation for the molecules of liquids in porous media can be analyzed with different models [13 - 16, 21, 22]. Callaghan *et al.* showed [16] that if the pores are small and have size r, then in the model of long-time limit, *i.e.*, at $\Delta >> r^2/D$, the attenuation of echo signal can be approximated by the expression:

$$\frac{I(G)}{I(0)} = \exp(-\beta^2 r^2) \tag{5}$$

where $\beta^2 = q^2/5$ and $q = \gamma G\delta$. Further, this expression was developed for an application to a Gaussian volume distribution of spherical pores [16, 21]:

$$\frac{I(\delta, G, r_0, \sigma)}{I(0)} = \frac{1}{\sqrt{1+2\beta^2\sigma^2}} \exp(-\frac{\beta^2 r_0^2}{1+2\beta^2\sigma^2}) \tag{6}$$

where σ is standard deviation, and r_0 is average size of pores. In white cement pastes, the T_1 and T_2 relaxation times are depending on pore size. Refs. [27, 28] showed that, in assumption of spherical pores, $T_1 = \alpha T_2$ and $T_2 = r/3\lambda$. Here, α is a constant, and λ is the surface relaxivity. Therefore, the integral formulae for a pore size distribution can be written as [2]:

$$I(G,r) = \int_r I_0(r)\exp(-\beta^2 r^2)\exp(-6\lambda\tau_2^*/r)\exp(-3\lambda\tau_1^*/\alpha r)dr \quad (7)$$

In the case of the spin-echo pulse sequence, τ_2* is the pulse interval between (π/2) and (π) radio-frequency (RF) pulses, and τ_1* is zero. In the stimulated echo experiment τ_1* is the gap between second and third RF pulses whereas τ_2* is the interval between first and second RF pulses. $\Delta = \tau_1^* + \tau_2^*$. The distribution of pore volumes is in proportional dependence to $I_0(r)$. The data can be acquired as a function of applied gradient G, τ_2* and Δ. The attenuation of echo signal is analyzed using the numerical integration and a total fit to the data at all experimental parameters and variables. Similar cement pastes were experimentally studied at frequency 20 MHz [28] resulting in $\alpha = 4$ and $\lambda = 0.0037$ nm/μs. These parameters can be frequency dependent. At 20 MHz, α is circa 9 times less than that at 400 MHz [27, 28]. However, λ is expected to be comparable for both 400 MHz and 20 MHz measurements [2, 8 and Refs. therein].

DDCOSY Experiments in Studying Anisotropy

Chapter 1 discussed some details of the 2D diffusion-diffusion correlation spectroscopy. Additional information with the presentation of DDCOSY data as 2D diffusion maps can be found in the literature [14, 15, 21, 29]. The data $I(q_1^2, q_2^2)$ for the attenuation of spin-echo signal with respect to the independent q_1^2 and q_2^2 variables are presented by eq. (8) [2, 14, 21, 29]:

$$\frac{I(q_1^2, q_2^2)}{I_0} = \exp(-q_1^2 D_1 \Delta_1)\exp(-q_2^2 D_2 \Delta_2) \quad (8)$$

The resultant 2D data set can be treated by 2D inverse Laplace transformation. The problem of 2D-inversion has been solved by Song *et al.* [30, 31]. They reduced the size of the 2D matrices using the Singular Value Decomposition. With this approach it was possible to apply the developed method to T_2-D correlation and to T_1-T_2 correlation experiments producing 2D spectra [32, 33]. In diffusion-diffusion correlation experiments 2D spectrum is created with the aid of double Laplace inversion applied to the echo - attenuation data sets from independent PFG NMR measurements with different orientation of gradients. DDCOSY experiment can investigate the local diffusion anisotropy in the case when sample is macroscopically isotropic. In literature there are examples of macroscopically isotropic systems with the locally anisotropic diffusion, *e.g.*, lamellar phase liquid crystals and strained elastomers [21, 29]. The PFG SE measurements in DDCOSY experiment are independent. However, DDCOSY correlates diffusional movement in two dimensions. In the suggestion of planar spaces in C-S-H gel, DDCOSY can also look for locally anisotropic motion within macroscopically isotropic hydrated cement samples.

The spectrum obtained in 2D diffusion map, can discover anisotropy of the diffusion tensor [15]. In the publications [21, 29, 34], the system with locally axially symmetric tensor of diffusion has been considered and analyzed in conditions of the powder average [21, 29]. According to the scheme presented in Ref. [34], the polar geometry of a locally anisotropic domain for diffusion motion is introduced. It is assumed that domain has cylindrical symmetry. These domains can be randomly oriented. The scheme [29] presents molecular frame based on (accepted in mathematics) a spherical coordinate system for three-dimensional space in which a point is characterized by radius-vector, polar ange θ and azimuth angle φ. As a molecular coordinate system is considered with respect to the laboratory coordinate system (*XYZ*), then a polar angle is measured from a fixed *Z*-direction to a radius-vector (director in molecular frame), and the azimuth angle is measured in reference plane (*XY*) from a fixed reference direction (*X*) to orthogonal projection of a radius-vector on the reference plane (*XY*).

In the assumption of locally anisotropic diffusion in molecular frame, the protons of water have the diffusion coefficient D_1 in the direction of radius-vector (this director is defined at polar angle θ and azimuth angle φ). Second diffusion coefficient D_2 characterizes a direction in the plane which is orthogonal to selected director (D_1). In such a local geometry [29, 34], the direction of applied gradient is defined by the polar axis. In DDCOSY experiments, echo signals are formed at the use of two gradient pairs which can be applied independently in different directions. If the first gradient pair is applied along *Z*-direction, whereas the second one is in the orthogonal *X*-direction, then the attenuation of spin-echo signal can be analytically presented by the eq. (9) [34]:

$$
\begin{aligned}
I(G_{1z}, G_{2x}) = & \int_0^1 d\cos\theta \exp[-q_{1z}^2 \Delta(D_1 \cos^2\theta + D_2 \sin^2\theta)] \\
& \times (2\pi)^{-1} \int_0^{2\pi} d\varphi \exp[-q_{2x}^2 \Delta(D_1 \sin^2\theta \cos^2\varphi + D_2 \sin^2\varphi + D_2 \cos^2\theta \cos^2\varphi)]
\end{aligned} \tag{9}
$$

Refs. [14, 15] presented simulated 2D diffusion-diffusion correlations with PFG parameters worked out at DDCOSY experiments with pairs of orthogonally oriented gradients in studying diffusion anisotropy in wood samples. Those calculations used eq. (9) to generate echo-attenuations and apply 2D ILT to 2D data sets. The 2D diffusion maps discovered diagonal peak with off-diagonal features (wings) as confirmation of anisotropic diffusion coefficients in simulated and experimental data [15].

COMPARISON OF PULSE SEQUENCES IN CEMENT STUDIES

In order to clarify an influence of internal gradients, the results of diffusion experiments in early hydration cement pastes have been compared for stimulated echo and 13-interval *Cotts* sequences. The results on the cement pastes with the hydration age of 2.5 and 9 days are shown in the Fig. (**2**) for diffusion time of 18 ms.

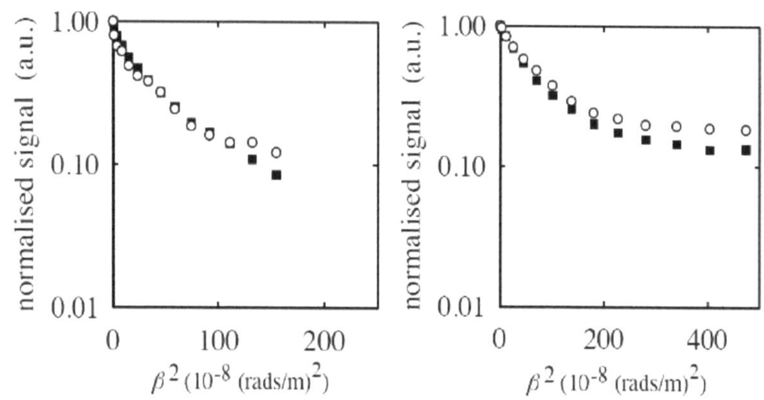

Fig. (2). The echo-attenuation curves for white cement paste with hydration time of 2.5 (left) and 9 (right) days. The diffusion time is 18 ms.
Open circles – the data from the *Cotts* sequence; *solid squares* – the data from the stimulated echo sequence.

Fig. (**3**) shows the comparison of the diffusion data (at diffusion time of 30 ms) from the *Cotts* and the stimulated echo sequences for cement paste with age of 4 days and for water. First of all, the data sets for water showed excellent agreement between two pulse sequences. This was expected because of water sample does not acquire additional internal gradients of magnetic field and should not result in any differences at applying STE and *Cotts* pulse sequences. The diffusion experiments on white cement pastes discovered also good agreement between these two pulse sequences. Heterogeneity of the magnet susceptibility is responsible for the appearance of background gradients. Internal gradient is seemed to have less effect than it was suggested before.

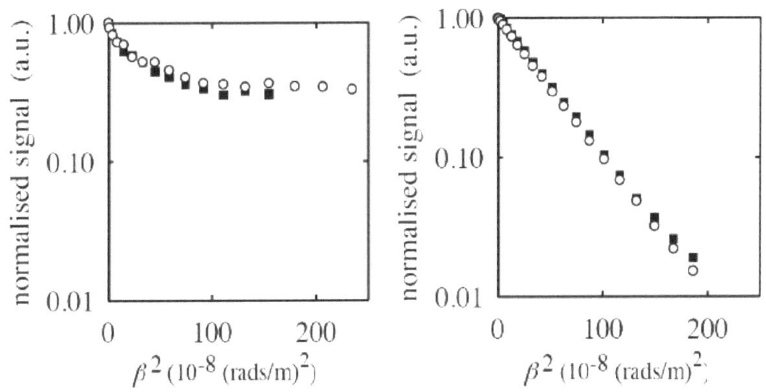

Fig. (3). The echo attenuation data from the stimulated echo (solid squares) and the *Cotts* (open circles) pulse sequences for white cement paste with hydration time of 4 days (left) and for water (right). The diffusion time is 30 ms.

The main source of paramagnetic impurities affecting magnetic susceptibility in cements is associated with iron. The amount of iron in white cement is very small in comparison with other cements. For example, the phase of ferrite (C_4AF) is about 7.1 % in grey C and 10.0% in grey B cements [8, 9, 35]. Iron probably is not uniformly distributed in the space of cement paste. According to [35], this iron in the paste rather aggregates in hydrogarnet phases or nano-crystalline AFm. That's probably why the internal gradients do not have strong effect.

Based on these comparative data (Fig. **2** and **3**), it was not necessary to study further a hydration of white cement paste using the *Cotts* pulse sequence. Therefore, the study [2] did not consider background gradients further and did not implement *Cotts* sequence additionally to the STE NMR diffusion measurements. It was noticed [2] that the introduction of additional gradients (in particular, bipolar gradients establish switching delays into the pulse sequence) increases signal relaxation attenuation, as well as amplifies other mechanisms resulting in additional signal loss. Wu *et al.* discussed these effects in details showing a limitation of the experiment with reduced S/N [36]. Pore size distributions were analyzed using eqs. (6) and (7).

SELF-DIFFUSION OF WATER IN CEMENT PASTE

Early Hydration of Cement Paste as Studied by 1-dimensional PFG NMR

The study of spin-spin relaxation times in white cement paste has shown that CPMG measurement is a function of cement age. Fig. (**4**) shows the corresponding T_2 distributions for cement paste after 15 hours, 2, 7 and 14 days of hydration. The data discovered that at the early stages of hydration, three T_2 peaks positioned at around 0.1, 0.4-0.6 and 2 ms are presented after applying ILT to CPMG data sets.

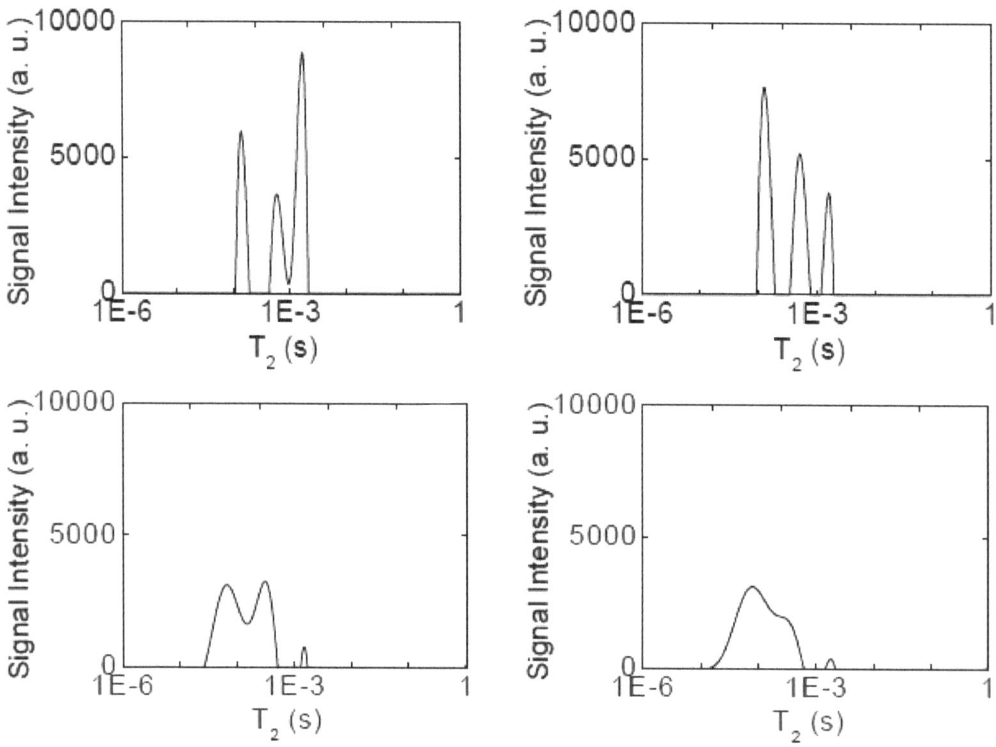

Fig. (4). The T_2 distribution for white cement paste with curing time of 15 hours (top left), 2 (top right), 7 (bottom left) and 14 (bottom right) days.

Careful analysis of the T_2 distributions and absolute signal intensities shows that the overwhelming majority of the signal arises from capillary network. During cement hydration from 15 hours to 14 days the signal with T_2 ~ 1-2 ms responsible for the micron sized pores with capillary water decreased from 50% to 1%. It is this fraction of water that determines the echo signal in NMR diffusion experiments.

It is hard to measure the data set with many parameters at the early cement hydration stages as the morphology of cement sample is evolving very quickly. Therefore, in Ref. [2], the measurements with sufficient signal-to-noise ratio have been made only for Δ=6 ms and δ =2 ms with a series of G values. The results of these measurements are presented in the Fig. (**5**) for different values of hydration time.

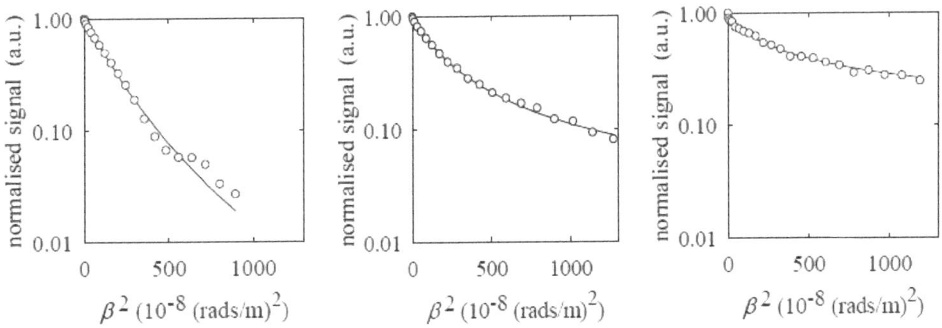

Fig. (5). Spin-echo-attenuation curves for white cement paste after various hydration times. $\Delta = 6$ ms, and $\delta = 2$ ms. *From left to right*: 15 hours, 2 days and 14 days of hydration. Solid lines are the fits of the echo-attenuation data according to eq. (6).

The data showed that in hydrated paste, the $\log(I(G)/I(0))$ is increasingly non-linear with β^2 (*i.e.* G^2) indicating restricted diffusion in a distribution of pore sizes, whereas it is linear for bulk water. The equation (6) has been used to fit the echo-attenuation curves producing an evaluation of the mean pore size. The analysis has discovered that the mean pore size is a function of hydration time. The pore size for early hydration stage (*i.e.* after 15 hours) was calculated as 8 μm. This was the largest mean pore size. After 7 days it becomes 4 μm and further keeps this value during two weeks of hydration. The data fits resulted in the pore size distribution widths which are comparable to the mean sizes. This suggests a broad distribution indeed. However, this reflects the limitations of the model too. Using the shortest experimental value $\Delta = 6$ ms and the diffusion coefficient of free water (which is 2.3×10^{-9} m^2/s [13, 14]) we can estimate within the long-time limit as $(6D\Delta)^{1/2} = 9$ μm, which is > 8 μm.

DDCOSY in Studying Early Cement Hydration

Earlier [14], we compared DDCOSY results studying anisotropic diffusion of water in wood (water in the wood is confined within highly anisotropic cells, discovering tens of microns in restricted size) and isotropic diffusion of bulk water. During DDCOSY studies of early hydration cement paste, the bulk water sample was also tested in parallel 2D NMR diffusion measurement with the same pulse sequence. The results on cements have been compared with 2D pattern for isotropic diffusion. The bulk water diffuses freely in all directions, and this is confirmed as one round spot on the main diagonal in resultant 2D diffusion-diffusion correlation map. These data and the data of 2D DDCOSY experiments on early hydration cement pastes are showed in Fig. (**6**).

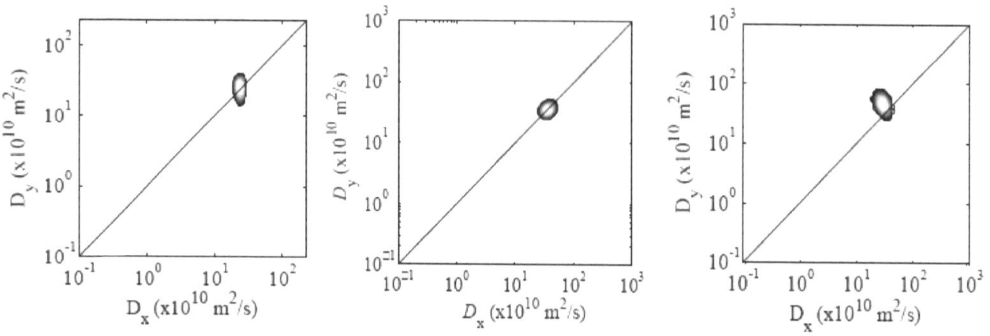

Fig. (6). 2D DDCOSY spectra for bulk water (*left*), and cement paste aged 1 day (*centre*) and 2 days (*right*). Each spectrum is derived from a 20×20 data set at $\delta_{1,2}$ = 1 ms, $\Delta_{1,2}$= 6 ms. G_{max}= is 1.2 T/m. The data in the centre (1 day) was recorded at 3 °C in order to slow hydration of the cement during measurement.

The 2D map for water is consistent with simulations of isotropic diffusion, published in [15, 20]. The findings of Fig. (**6**) (*centre* and *right*) discovered also that 2D DDCOSY spectra of 1 and 2 days cements have the only spot on the diagonal. The 2D spectrum for cement paste after 7 days of hydration has discovered the same features with the only diagonal spot. 2D diffusion experiment is not fast in comparison with the hydration kinetics at early stage of cement paste hardening. In order to slow the hydration process in cement sample aged 1 day and to carry out the measurement with an unchanging sample during the acquisition time of 2D diffusion experiment, the temperature of the cement paste was hold 3 °C. The data of the plots (Fig. (**6**), *centre* and *right*) are similar to the finding of Fig. (**6**) (*left*) for isotropic diffusion of bulk water and suggest isotropic water diffusion in early capillary network. Thus, DDCOSY did not discover any diffusion anisotropy in the cement pores. This is also in line with other publications [37, 38] showing anisotropy of small gel pores, and absence of anisotropy in the capillary pores.

Pore Size Distribution and Self-diffusion of Water in Mature Cement Paste

Spin-spin relaxation time in cements is strongly pore size dependent [8, 28]. Though the average size of cement pores is about few microns, the pores of large sizes can be over-represented in total echo signal. Some details of pore size distribution could be clarified in additional study of NMR diffusion in mature cements doing the experiments at various diffusion times Δ. The hydration changes in the cement paste with large age are mostly slow, and this allow to acquire the multiparameter data set for cement samples even in the experiments with relatively long measurement time, for example, studying diffusion coefficient as a function of diffusion time. Using SE and STE pulse sequences, a data set was obtained for mature cement paste (age is about 1 year) as a function of diffusion time Δ in the range from 6 ms to 50 ms at $\delta = 2$ ms. The applied gradient was varied from 0 to 1.2 T/m in order to get the dependence of echo-attenuation *vs* G^2 [2]. It was suggested that a distribution of pore sizes in mature cement paste obeys to the log-normal volume distribution, according to [2]:

$$I_V(r) \propto \frac{1}{\sigma r\sqrt{2\pi}} \exp\left(-\frac{(\ln(r/r_0))^2}{2\sigma^2}\right) \quad \textbf{(10)}$$

Here, the mean pore size is r_0. Width parameter σ is dimensionless.

On the basis of this distribution, the equation (7) has been used to make a total fit to the data with $\alpha = 35$ [2]. Fig. (7) shows the solid lines as the results of the fit to the experimental data at the diffusion times from 6 (top) to 50 ms (bottom). The findings from the fitting: $\sigma = 0.51$, $\lambda = 0.7$ nm/µs, $r_0 = 4.2$ µm.

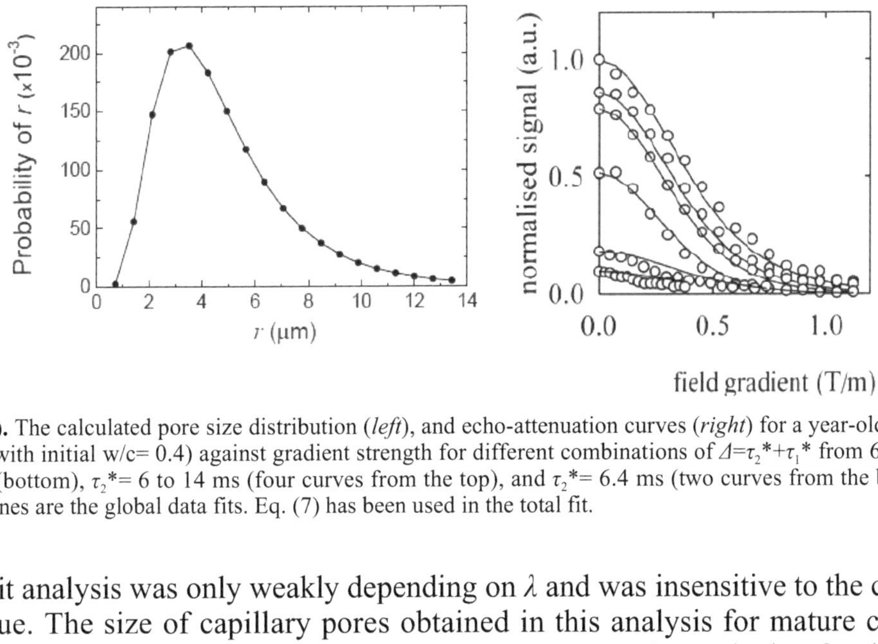

Fig. (7). The calculated pore size distribution (*left*), and echo-attenuation curves (*right*) for a year-old cement paste (with initial w/c= 0.4) against gradient strength for different combinations of $\Delta=\tau_2^*+\tau_1^*$ from 6 (top) to 50 ms (bottom), τ_2^*= 6 to 14 ms (four curves from the top), and τ_2^*= 6.4 ms (two curves from the bottom). Solid lines are the global data fits. Eq. (7) has been used in the total fit.

The fit analysis was only weakly depending on λ and was insensitive to the chosen α-value. The size of capillary pores obtained in this analysis for mature cement paste is comparable to results for early hydration pastes, in particular, for the size of pores after seven days of hydration. These results suggest that hydrates and gel porosity do not form in the capillary porosity once the latter has been substantially created.

In order to estimate T_2 and T_1 values, it is possible to fit the plots of ln(I) at G=0 against $2\tau_2^*$ in spin-echo pulse data or against τ_1^* at constant τ_2^* for the data from stimulated echo sequence. With this fitting, T_2 and T_1 values have been found as

24 and 43 ms, respectively. Although the values of T_2 and T_1 look less than it was predicted by scaling low frequency, however, for this analysis, they are quite reasonable. The discrepancies could be associated with the simplified application of the relaxation model in large pores. It is also possible to attribute the effect to residual diffusive attenuation because of background gradients. Other methods to measure T_2 (*e.g.* using echo trains with short spaced echoes) can clarify the issue.

CONCLUDING REMARKS

The chapter reported one- and two-dimensional ^{1}H PFG studies of water in white cement pastes with w/c = 0.4 cured for periods of one to fourteen days and one year. In more mature samples, the signal arises almost exclusively from the small fraction of residual capillary water. The capillary water is found to be in confining environments. A well known expression for restricted diffusion in a Gaussian distribution of pore sizes has been adapted to the case of log-normal distribution for which the relaxation times T_1 and T_2 are pore size dependent. Data for the various combinations of Δ (from 6 to 50 ms), and for the different RF pulse gaps in the spin-echo and stimulated echo pulse sequences were considered. A global fit of all diffusion data together (on the basis of a log-normal pore size distribution) results in the volume mean capillary pore size 4.2 μm in mature paste, that is similar to pore size at 7-14 days of hydration. In order to negate the effects of internal magnetic field susceptibility gradients in hardening cement pastes, the 13-interval *Cotts* pulse sequence with bipolar gradient pairs has been implemented and applied in early hydration cement pastes (2-9 days). It was shown that the effect of internal gradients was rather small. 2D DDCOSY experiments with two pair of orthogonal gradients have been applied first time to hydrated cement pastes. Two-dimensional *D-D* correlation experiments have been carried out in cement pastes at room temperature and at 3 °C exploring water diffusion and nascent pore anisotropy during the early stages of hydration. No evidence was found of capillary pore anisotropy in cement paste.

ACKNOWLEDGEMENTS

The author thanks Peter McDonald.

REFERENCES

[1] I. Maruyama, N. Sakamoto, K. Matsui, and G. Igarashi, "Microstructural changes in white Portland cement paste under the first drying process evaluated by WAXS, SAXS, and USAXS", *Cement Concr. Res.*, vol. 91, pp. 24-32, 2017. [http://dx.doi.org/10.1016/j.cemconres.2016.10.002]

[2] V.V. Rodin, P.J. McDonald, and S. Zamani, "A nuclear magnetic resonance pulsed field gradient study of self-diffusion of water in hydrated cement pastes", *Diffus. Fundam.*, vol. 18, no. 3, pp. 1-7, 2013.

[3] V.N. Izmailova, V.V. Rodin, E.D. Shchukin, G.P. Yampolskaya, P.V. Nuss, A.N. Ovchinnikov, and Z.D. Tulovskaya, NMR-approach in the study of the influence of nonionic surfactant on water state during cement hardening.*Nuclear Magnetic Resonance Spectroscopy of Cement-Based Materials.*, P. Colombet, A-R. Grimmer, H. Zanni, P. Sozanni, Eds., Springer-Verlag-Press, 1998, pp. 157-165. [http://dx.doi.org/10.1007/978-3-642-80432-8_28]

[4] S. Rahimi-Aghdam, Z.P. Bažant, and M.J. Abdolhosseini Qomi, "Cement hydration from hours to centuries controlled by diffusion through barrier shells of C-S-H", *J. Mech. Phys. Solids,* vol. 99, pp. 211-224, 2017. [http://dx.doi.org/10.1016/j.jmps.2016.10.010]

[5] A. Pop, C. Badea, and I. Ardelean, "The effects of different superplasticizers and water-to-cement ratios on the hydration of gray cement using T_2 NMR", *Appl. Magn. Reson.,* vol. 44, no. 10, pp. 1223-1234, 2013. [http://dx.doi.org/10.1007/s00723-013-0475-5]

[6] V.N. Izmailova, V.V. Rodin, E.D. Shchukin, G.P. Yampolskaya, and P.V. Nuss, "The effect of nonionic surfactants on the state of water in cement systems (by NMR relaxation data): 1. The state of water in the course of structure formation", *Colloid J.,* vol. 60, no. 1, pp. 5-12, 1998.

[7] V.N. Izmailova, V.V. Rodin, E.D. Shchukin, G.P. Yampolskaya, and P.V. Nuss, "The effect of nonionic surfactants on the state of water in cement systems (by NMR relaxation data). 2. A model of pore space", *Colloid J.,* vol. 60, no. 1, pp. 13-18, 1998.

[8] A. Valori, V.V. Rodin, and P.J. McDonald, "On the interpretation of ^1H 2-dimensional NMR relaxation exchange spectra in cements: is there exchange between pores with two characteristic sizes or Fe^{3+} concentrations?", *Cement Concr. Res.,* vol. 40, no. 9, pp. 1375-1377, 2010. [http://dx.doi.org/10.1016/j.cemconres.2010.03.022]

[9] V.V. Rodin, A. Valori, and P.J. McDonald, "A ^1H double quantum filtered NMR study of water in cement pastes", *New J. Physics,* vol. 13, p. 035017 (app 14), 2011.

[10] C. Badea, A. Pop, C. Mattea, S. Stapf, and I. Ardelean, "The effect of curing temperature on early hydration of gray cement *via* fast field cycling-NMR relaxometry", *Appl. Magn. Reson.,* vol. 45, p. 1299, 2014. [http://dx.doi.org/10.1007/s00723-014-0565-z]

[11] P.S. Wang, M.M. Ferguson, G. Eng, D.P. Bentz, C.F. Ferraris, and J.R. Clifton, "^1H nuclear magnetic resonance characterization of Portland cement: molecular diffusion of water studied by spin relaxation and relaxation time-weighted imaging", *J. Mater. Sci.,* vol. 33, pp. 3065-3071, 1998. [http://dx.doi.org/10.1023/A:1004331403418]

[12] N. Nestle, P. Galvosas, and J. Kärger, "Liquid-phase self-diffusion in hydrating cement pastes — results from NMR studies and perspectives for further research", *Cement Concr. Res.,* vol. 37, pp. 398-413, 2007. [http://dx.doi.org/10.1016/j.cemconres.2006.02.004]

[13] V.V. Rodin, and T. Cosgrove, "Nuclear magnetic resonance study of water-polymer interactions and self-diffusion of water in polymer films", *OALJ: Chem. Mater. Sci,* vol. 3, no. 10, pp. 1-17, 2016.

[14] V.V. Rodin, "Methods of Magnetic Resonance in Studying Natural Biomaterials", In: *Encyclopedia of Physical Organic Chemistry,* Zerong Wang, Ed., Chapter 53. vol. 4, part 4. John Wiley & Sons, Inc., 2017, pp. 2861-2908.

[15] V.V. Rodin, P.J. McDonald, and M. Jones, "Two-dimensional distribution function of diffusion in wood obtained using 2D Laplace inversion", *Appl. Phys. Math.,* no. 6, pp. 3-7, 2014.

[16] P.T. Callaghan, K.W. Jolley, and R.S. Humphrey, "Diffusion of fat and water in cheese as studied by pulsed field gradient nuclear magnetic resonance", *J. Colloid Interface Sci.,* vol. 93, no. 2, pp. 521-529, 1983. [http://dx.doi.org/10.1016/0021-9797(83)90436-8]

[17] R.F. Feldman, and P.J. Sereda, "A new model for hydrated Portland cement and its practical implications", *Engin. J. Canada,* vol. 53, no. 8-9, pp. 53-59, 1970.

[18] A.J. Allen, J.J. Thomas, and H.M. Jennings, "Composition and density of nanoscale calcium-silicate-hydrate in cement", *Nat. Mater.,* vol. 6, no. 4, pp. 311-316, 2007. [http://dx.doi.org/10.1038/nmat1871] [PMID: 17384634]

[19] R.M. Cotts, M.J.R Hoch, T. Sun, and J.T. Markert, "Pulsed field gradient stimulated echo method for improved NMR diffusion measurements in heterogeneous system", *J. Magn. Reson.,* vol. 83, no. 2, pp. 252-266, 1989.

[20] V.V. Rodin, and V.A. Nikerov, "NMR relaxation and PFG NMR studies of water dynamics in oriented collagen fibres with different degree of cross-linking", *Curr. Tissue Eng.,* vol. 3, no. 1, pp. 47-61, 2014. [http://dx.doi.org/http://dx.doi.org/10.2174/2211542003666140626211652]

[21] P.T. Callaghan, *Translational Dynamics and Magnetic Resonance.* Oxford University Press: Oxford, 2011. [http://dx.doi.org/10.1093/acprof:oso/9780199556984.001.0001]

[22] J.E. Tanner, and E. Stejskal, "Restricted self-diffusion of protons in colloidal systems by the pulsed-gradient, spin-echo method", *J. Chem. Phys.,* vol. 49, no. 4, pp. 1768-1777, 1968. [http://dx.doi.org/10.1063/1.1670306]

[23] J.E. Tanner, "Transient diffusion in system partitioned by permeable barriers. Application to NMR measurements with a pulsed field gradient", *J. Chem. Phys.,* vol. 69, no. 4, pp. 1748-1754, 1978. [http://dx.doi.org/10.1063/1.436751]

[24] J.E. Tanner, "Use of the stimulated echo in NMR diffusion studies", *J. Chem. Phys.,* vol. 52, no. 5, pp. 2523-2526, 1970. [http://dx.doi.org/10.1063/1.1673336]

[25] E.O. Stejskal, and J.E. Tanner, "Spin diffusion measurements: Spin echos in the presence of a time-dependent field gradient", *J. Chem. Phys.,* vol. 42, no. 1, pp. 288-292, 1965. [http://dx.doi.org/10.1063/1.1695690]

[26] R.F. Karlicek, and I.J. Lowe, "A modified pulsed gradient technique for measuring diffusion in the presence of large background gradients", *J. Magn. Reson.,* vol. 37, no. 1, pp. 75-91, 1980.

[27] K.S. Mendelson, W.P. Halperin, J-Y. Jehng, and Y-Q. Song, "Surface magnetic relaxation in cement pastes", *Magn. Reson. Imaging,* vol. 12, no. 2, pp. 207-208, 1994. [http://dx.doi.org/10.1016/0730-725X(94)91517-2] [PMID: 8170300]

[28] L. Monteilhet, J.P. Korb, J. Mitchell, and P.J. McDonald, "Observation of exchange of micropore water in cement pastes by two-dimensional T_2-T_2 nuclear magnetic resonance relaxometry", *Phys. Rev. E.,* vol. 74, no. 6, p. 061404, 2006. [http://dx.doi.org/10.1103/PhysRevE.74.061404] [PMID: 17280070]

[29] P.T. Callaghan, and I. Furó, "Diffusion-diffusion correlation and exchange as a signature for local order and dynamics", *J. Chem. Phys.,* vol. 120, no. 8, pp. 4032-4038, 2004. [http://dx.doi.org/10.1063/1.1642604] [PMID: 15268569]

[30] Y.Q. Song, L. Venkataramanan, M.D. Hürlimann, M. Flaum, P. Frulla, and C. Straley, " T_1-T_2 correlation spectra obtained using a fast two-dimensional Laplace inversion", *J. Magn. Reson.,* vol. 154, no. 2, pp. 261-268, 2002. [http://dx.doi.org/10.1006/jmre.2001.2474] [PMID: 11846583]

[31] L. Venkataramanan, Yi-Q. Song, and M.D. Hurlimann, "Solving Fredholm integrals of the first kind with tensor product structure in 2 and 2.5 dimensions", *IEEE Trans. Signal Process.,* vol. 50, no. 5, pp. 1017-1026, 2002. [http://dx.doi.org/10.1109/78.995059]

[32] M.D. Hürlimann, and L. Venkataramanan, "Quantitative measurement of two-dimensional distribution functions of diffusion and relaxation in grossly inhomogeneous fields", *J. Magn. Reson.*, vol. 157, no. 1, pp. 31-42, 2002.
[http://dx.doi.org/10.1006/jmre.2002.2567] [PMID: 12202130]

[33] R.G. Graham, W.M. Holmes, C. De Panfilis, and K.J. Packer, "Characterisation of locally anisotropic structures within isotropic porous solids using 2D pulsed field gradient NMR", *Chem. Phys. Lett.*, vol. 332, pp. 319-323, 2000.
[http://dx.doi.org/10.1016/S0009-2614(00)01279-3]

[34] P.T. Callaghan, "How two pairs of gradient pulses give access to new information about molecular dynamics", *Diffus. Fundam.*, vol. 2, pp. 64.1-64.18, 2005.

[35] B.Z. Dilnesa, *"Fe-containing hydrates and their fate during cement hydration: thermodynamic data and experimental study"*, PhD thesis, EPFL, Switzerland, 2011.

[36] D.H. Wu, A.D. Chen, and C.S. Johnson, "An improved diffusion-ordered spectroscopy experiment incorporating bipolar-gradient pulses", *J. Magn. Reson. A,* vol. 115, pp. 260-264, 1995.
[http://dx.doi.org/10.1006/jmra.1995.1176]

[37] E. Gallucci, K. Scrivener, A. Groso, M. Stampanoni, and G. Margaritondo, "3D experimental investigation of the microstructure of cement pastes using synchrotron X-ray microtomography (μCT)", *Cement Concr. Res.*, vol. 37, pp. 360-368, 2007.
[http://dx.doi.org/10.1016/j.cemconres.2006.10.012]

[38] I.G. Richardson, "The nature of C-S-H in hardened cements", *Cement Concr. Res.*, vol. 29, pp. 1131-1147, 1999.
[http://dx.doi.org/10.1016/S0008-8846(99)00168-4]

CHAPTER 11

Characterisation of Hydrated Cement Pastes by ^{1}H DQF NMR Spectroscopy

Abstract: This chapter considers how ^{1}H DQF NMR spectroscopy characterises hydrated cement pastes. The increase of the DQF NMR signal during cement hydration has been observed. The mobile single-quantum (SQ) signal was decreasing with increasing hydration time. The measured DQF NMR signal in hydrated white cement has been fitted by a sum of two components. The protons of $Ca(OH)_2$ were responsible for the appearance of the first component. The second component was associated with the protons of water in the planar C–S–H gel pores. A model of water molecules movement near the centres of paramagnetic impurities has been considered. The model was consistent with the level of paramagnetic iron content and experimental data. The DQF spectra in grey cement pastes have been fitted by a sum of three components. First two components in grey cement were considered similar to white cement. The iron-rich phases in grey cements were responsible for an appearance of third component. The experiments with progressively heating cements showed how water leaves from cement sample and how the DQF signals of all components changed with removing water. The behaviour of the DQF signal as function of relative sample mass was in sympathy with bound component of the solid echo experiment. It was discovered that the DQF component assigned to the C–S–H water decreases monotonically, while that associated with the solid $Ca(OH)_2$ first increases before decreasing.

Keywords: $Ca(OH)_2$, Creation time, Double-quantum-filter (DQF) NMR, Evolution time, Fourier transform (FT), Free induction decay (FID), Iron content, NMR spectra, Residual dipolar interaction (RDI), Spin-lattice relaxation time T_1, Spin-spin relaxation time T_2, White and grey cement pastes, Water.

1D AND 2D ^{1}H NMR RELAXATION

The hardening of cement is a hydration phenomenon [1]. When cement contacts with water, the highly soluble components of cement powder dissolve quickly creating a supersaturated solution with hydrates components. This causes precipitation. The hydrates grow on the surface of the cement grain. During the cement hydration, there is a chemical shrinkage. Therefore, cement is intrinsically and inevitably porous.

Victor V. Rodin

The macro properties of cement pastes are defined by cement porosity and pore-water interactions. Different cement pastes are used in production of concrete. Concrete is an inherent heterogeneous material. Its performance depends on the properties of water in hardening the cement paste [1 - 3]. The dynamics of water confined within hydrated cement structure [2 - 4] affects the degradation of concrete. In order to develop new models describing the cement hardening and to improve the properties of new cement based materials, it is necessary to study the detailed behaviour of water in cement pastes.

Cement has its own chemistry nomenclature. All elements in cement powder are described in terms of oxides. For example, calcium oxide (CaO) has cement notation formula: C, silica (SiO_2) has cement notation formula: S, aluminium oxide (Al_2O_3) has cement notation formula: A, and iron oxide (Fe_2O_3) has cement notation formula: F. Cement powder is the combination of oxides of mostly calcium, silicon, aluminium and iron. Using the cement notation, the main components of Portland cement are C_3S and C_2S (called Alite and Belite). The two phases that develop the main strength during hydration are a gel like calcium silicate hydrates (C-S-H) and calcium hydroxide ($Ca(OH)_2$ or CH or Portlandite) [1, 3, 5]. The main reaction in the hydration of cement is the hydration of the C_3S and C_2S. The reaction products can be subdivided into different phases, each one having its own structure. The pores can be divided in to capillary and gel pores on the basis of size formation and location [4 - 6]. Water is considered to be part of the structure of hydrated cement since it plays a key role in many processes, from hardening to degradation.

Several NMR techniques have been used in studying porous media to measure different physical parameters [2, 4 - 12]. The protons of liquid water in hardening cement paste can be observable with NMR spectroscopy. The bulk water has a narrow NMR spectrum with Lorentzian like shape and the width at half height of one kHz. The width of proton spectra of solid $Ca(OH)_2$ is about 60 kHz. The resonance lines of the solid and liquid protons in cement paste overlap [11, 12]. Free bulk water signal is normally very intensive and can mask the proton contribution of the solid matrix. So, it is not easy to study the protons of the solid cement product, in particular, the C-S-H gel and calcium hydroxide $Ca(OH)_2$. Therefore, some approaches of 2D NMR were developed to study these products in the cement hydration process [4, 7, 12]. An essential role in detailing the cement paste properties and developing the new models on water dynamics in the pore structure is associated with different experiments in NMR relaxometry [2, 4 - 9, 11]. T_1 and T_2 relaxation times result in the information related to the proton dynamics. Bloembergen-Purcell-Pound (BPP) theory [13] is probably the base to be developed further. The theory describes the spectral density functions of the motion of the spins in the neighbours of paramagnetic centres (*i.e.*, in a strong internal magnetic field) and uses these to derive the relaxation times. NMR is suitable to study porous media because there is an increase of the relaxation rate of the liquid when introduced in the empty spaces of the material compared to when the liquid is in bulk [14 - 19]. This must be related to interaction at the solid-liquid interface. Some assumptions should be hold to apply the theory for many applications [16 - 22].

These assumptions are the short-range surface interactions and a fast exchange of molecules between the surface region and the bulk fraction of the liquid. Then, that allows averaging between the two environments. At these assumptions, the magnetization decays exponentially with a relaxation rate that is the weighted average between the surface and the bulk rates [5, 14, 15].

$$\frac{1}{T_{obs}} = \frac{1}{T_b}\left(1 - \frac{\lambda S}{V}\right) - \frac{1}{T_s}\frac{\lambda S}{V} \qquad \textbf{(1)}$$

In eq. (1), T_b and T_s are the bulk and surface relaxation times, V and S are the volume and surface area of the pore, respectively. λ is the thickness of a surface layer. Value T is spin-spin (T_2), or spin-lattice (T_1) relaxation time. According to [14], it is possible to present eq. (1) easier if to introduce the strength of the surface relaxation ρ as:

$$\rho = \lambda\left(\frac{1}{T_s} - \frac{1}{T_b}\right) \qquad \textbf{(2)}$$

ρ is the physical constant that does not depend on the pore size. In the most cases, $T_b >> T_s$. Then, the eq. (1) can be presented (applying to the case of spin-spin relaxation rate $1/T_2$) as follows [14, 15]:

$$\frac{1}{T_{2obs}} - \frac{1}{T_{2b}} = \frac{\rho S}{V} \qquad \textbf{(3)}$$

It is possible to consider the observed magnetization as a sum of magnetization decays for all pores. Then, the total magnetization separates in the components with the amplitudes of each component. According to [14], the echo intensity measured at a time t for the pore with the volume V and surface area S is following to eq. (4):

$$M_t = M_0 \exp\left[-\frac{\lambda}{T_{2s}}\frac{S}{V}t\right] \quad \textbf{(4)}$$

Cement paste is an inhomogeneous system with the pores filled by water that is in an intermediate/slow exchange indeed on the NMR experiment time scale [14]. The different areas of porous cement paste have different *S*/*V* ratios. The water molecules relax with different rates in these areas. Then, the observed NMR signal is the superposition of the signals from the different areas [15 - 17]. We consider that there is a distribution of the pore sizes in cement sample. According to [14], a total signal is presented by eq. (5):

$$M_t = M_0 \int P(r) \exp\left[-\frac{\lambda}{T_{2s}}\frac{S}{V}t\right] dr \quad \textbf{(5)}$$

Here, *P*(*r*) is the function of a pore volume distribution that obtained from the measured magnetisation M_t. The thickness of the surface layer λ can be considered as 3 Å.

From T_2 measurement, it is possible to determine a pore size distribution [5]. For example, following the Laplace inversion [18], a decay of spin-echoes train (registered in CPMG experiment) converted into T_2 distribution (Fig. **1**). After finding the spin-spin relaxation times distribution, further, eq. (3) and eq. (5) could be applied. The application of these equations leads to the pore size distribution that has the same shape as the T_2 distribution, due to the linearity of eq. (3). Thus, ^{1}H NMR relaxation studies have resulted in evidence of a distribution of pore types in hydrated cement [4, 6 - 9]. The lack in knowledge of the relaxation times for the surface layer can be main limitation to apply the equations (3) and (5) successfully for a determination of the absolute pore sizes.

One-dimensional relaxation CPMG measurements can be useful in determining the T_2 distribution function of liquids filling the empty spaces in the porous media. When the information about the T_2 distribution is known, then the pore size distribution could be got from the T_2 data obtained. However, this NMR approach gives a snapshot of the porosity without any information on the connectivity of the pores. 2D NMR relaxation analysis can help in determining the connectivity of the pores [4 - 8, 17 - 19].

For a sample with broad distribution of T_1 and T_2, the T_1-T_2 correlation spectra can better determine the T_1/T_2 ratio for each component compared to the separate T_1

and T_2 experiments [5, 16, 17]. In other cases, when there are spin populations, which have the same T_1 but different T_2, the issue can be solved with a 2D experiment but not with two 1D experiments. In general, a pulse sequence of any two-dimensional experiment can be separated into two parts. Then, the different Hamiltonians describe the processes in these parts of pulse experiment. The T_1-T_2 correlation experiment can be performed using an inversion recovery followed by a CPMG pulse sequence [4, 5, 18]. In this way, the experiment yields a two-dimensional n by m data set encoding T_1 in the first dimension and T_2 in the second one [5]. 2D NMR T_2-T_2 experiments can be used for studying an exchange [4, 7].

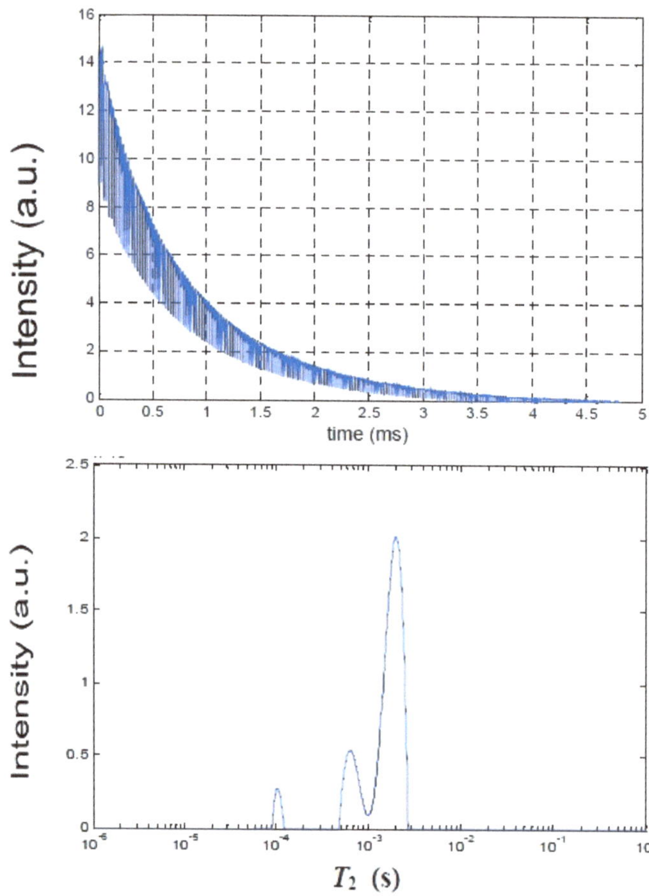

Fig. (1). The plots showing the steps in determining the pore size distribution from a CPMG (T_2) experiment (400 MHz, room temperature). *Top*: the experimental CPMG-decay measured in the hydrated cement (~50 min hydration in NMR tube, w/c=0.4). The intensities normalized per 10^5.
Bottom: T_2 distribution obtained with the inverse Laplace transformation [6]. The intensities (Y-axis) normalized per 10^4.

In the case of the diffusion of water molecules within the pore, the exchange is mostly fast. When a porous media has the pores of different sizes, the exchange should be considered between them. The fast exchange approximation will be not suitable for estimation of interporous exchange. Then, an intermediate exchange rate condition should be considered to describe the total magnetization, which will be the sum of the magnetisation signals of the two reservoirs, and they will influence each other [4, 5, 21].

The solutions of the system of basic equations governing the process of exchange during relaxation [5, 18, 19, 21, 22] are derived from the work of McConnell [23] and presented, for example, for T_2-T_2 experiment in [4]. The total result of 2D NMR correlation experiment is a relaxation spectrum with the four peaks. The two peaks are located on the diagonal, and other two ones are outside the diagonal. For the conditions of slow exchange, just two peaks are observed. In condition of very fast exchange only a single peak can be registered at the weighted average location. Comparative analysis of the T_1-T_2 and the T_2-T_2 experiments shows several advantages in using the T_2-T_2 correlation. The water exchange in the T_2-T_2 experiment is unambiguous. The T_2-T_2 spectrum has both peaks as positive [4]. In the case of the T_1-T_2 spectrum, one of the off diagonal peaks is negative [5]. An advantage of the T_2-T_2 experiment compared with the T_1-T_2 is also that it allows a quantitative estimation of the exchange rate *via* the third parameter of the experiment (the exchange time τ_{exch}). The analysis of the T_2-T_2 spectra at varying the exchange time was considered on simulated data in [4] showing that the cross peaks are observable only when the exchange time is comparable with the reciprocal of the exchange rate between the two pores. It has been found also that the different populations are affected by the increasing storage time in different ways.

Fig. (**2**) shows the example of T_2-T_2 spectrum obtained in correlation experiment on synthetic C-S-H. The plot discovered a strong diagonal peak (at T_2 ~20-30 ms) and a weak peak (at T_2 ~2 ms). The two exchange off-diagonal peaks are clearly visible. This is a typical relaxation pattern of two dominant reservoirs which are characterised also by two exchange peaks. More details and examples with exchange peaks observed in grey and white cements, C-S-H, and C_3S samples are in Refs. [4 - 6]. 2D ^{1}H NMR relaxation data have shown water exchange between pores with characteristic sizes of the order of a few nanometres and a few tens of nanometres, *i.e.*, between compartments which have the distinctive sizes [7]. First, the diagonal peaks (*e.g.*, as in Fig. **2**) inform about two pore sizes. At the same time, the appearance of the off-diagonal peaks is the result of water exchange between these pores [5]. The relaxation peaks in 2D T_2-T_2 correlation spectrum of the synthesised C-S-H are shifted to slower rates (T_2 value is circa by order higher) in comparison with white cement [7]. Electron spin resonance discovered that the paramagnetic concentration (iron content) in synthesised C-S-H is negligible as compared to white cement. The Fe^{3+} content in white cement is much lower than that in grey (B and C) cements [19]. The synthesised C-S-H is practically free of paramagnetic impurities. Much slower spin-spin relaxation rates in the synthesised C-S-H, *i.e.*, movement of the peaks up to the diagonal indicated a decreased surface relaxivity due to low concentration of paramagnetic iron (Fig. **2**) [7].

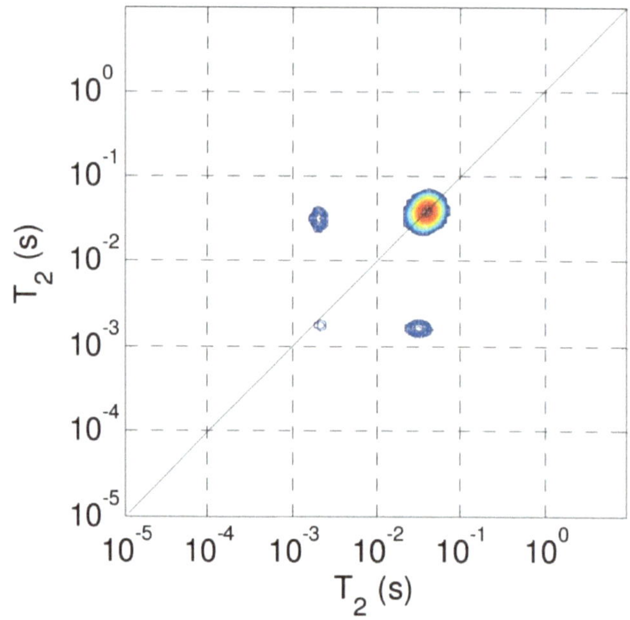

Fig. (2). A T_2-T_2 spectrum recorded in the experiment performed on synthesised C-S-H with storage time of 10 ms, the regularisation parameter used is α=10. A strong diagonal peak at T_2 value of ~30 ms and a much weaker one at ~2 ms are visible. S/N > 400.

Probably the complete deficiency of iron in the C-S-H is the first reason to register the peaks at much longer T_2 values (the considerable shift to longer relaxation times). Synthesised C-S-H has been prepared in suspension [5]. This could be the second reason because the particles are much more coarsely packed than in a hydrated cement paste. Then, it is possible to imagine the synthesis of the C-S-H as a hydration at a very high water-to-cement ratio. This gives a sort of Power model, a much higher porosity.

Pore-water interactions, a distribution of water, and an exchange between pores in hydrated cements considered until now are mainly based on the data revealed by NMR relaxometry. Water signal in the smallest pores could arise from the interlayer water in the C-S-H nanoparticles. The water in the spaces between the particles could be considered as next to the reservoir characterising the pores of other type and size. Probably, NMR relaxometry results are not enough to detail with water distribution further in hardening the cement pastes. The analysis of solid echo experiments and double-quantum-filtered (DQF) NMR spectroscopy data can discover the new evidences in correct assignment of the revealed water reservoirs.

Solid echo experiments show careful examination of the amplitude of the signal as a function of the mass [8]. This led to the formalisation of a simple drying model for the water restricted in the cement gel pores. The model is also capable to estimate the layers thicknesses. The results obtained are comparable with the T_2-T_2 estimations. In addition to this [8], the DQF NMR experiments discovered anisotropy of water distribution in cements and did show that the anisotropy of water increases as the sample is dried. According to earlier information discussed before in the literature, gel pores in cements are often considered as planar [24]. So, the DQF NMR results increased the confidence that the protons detected can be referred to the water of inter and intra C-S-H layers.

The next section considers the results of ^{1}H DQF NMR exploration of white and grey cement pastes [25, 26] detailing the water interactions in the pores of the different sizes. Multiple-quantum-filter (MQF) NMR techniques are additional approaches to standard single-quantum (SQ) NMR spectroscopy [27 - 29]. In particular, DQF method is distinct from SQ NMR. DQF NMR technique is sensitive to slow molecular dynamics in the systems with anisotropically confined and bound water [26, 27]. SQ NMR is a little informative method for studying these systems. The single excitation pulse is applied in SQ NMR. The signal is produced from all the nuclei of the sample as a response to this pulse. MQF NMR spectroscopy is the response to multiple pulse excitations. It works with a magnetization that has evolved coherently between the nuclei which are strongly coupled [28 - 32]. When isotropic molecular tumbling has a place, then magnetic dipolar coupling is averaging to zero. In contrast, an anisotropic molecular tumbling does not average all dipolar interactions to zero. The residual dipolar interactions (RDIs) between ^{1}H of water protons and solid protons result in second-rank tensor to be formed. Therefore, observed DQF NMR spectrum is due to the formation of second-rank tensor, and this is an indicator of anisotropic phases [18, 27]. DQF NMR signal is not created in the isotropic systems. That is why the DQF technique could be very useful in anisotropic planar pores studies to unveil pore structure.

In order to understand clearly pore morphology, it is necessary to estimate the pore width and the extent of anisotropic movement of water in the planar pores. At the exploration of porous water, the special NMR experiments should be involved unveiling a low mobility of the water protons. In addition, the properties of the bound water in the solid phases of hydrated cement (for example, in $Ca(OH)_2$) may differ from those of strongly adsorbed water in the narrowest pores of the C–S–H gel. DQF data could clarify these differences to assure or supplement the current interpretation of the discussed above NMR relaxation data [25]. The DQF NMR spectroscopy suppress the water signals in mobile regions, and only the proton signals originated in confined environments with bound and strongly anisotropic water are registered [26].

The chapter analyses further the ^{1}H DQF NMR study of hydration and cure of cement pastes showing the signal growth with hydration time. This was in line with the loss of relaxation (T_2) components measured in SQ experiments. The results obtained show also how the DQF signal of a cured cement paste varied in amplitude during progressively drying the sample. During this drying, the line shape of the DQF spectra was explored for different materials: pure $Ca(OH)_2$, white cement and two grey (B and C) cements with different iron content. The different components of line shape were assigned to hydrogen in different environments. The DQF data considered resulted in a model that water protons in the planar cement pores relaxed by Fe^{3+} paramagnetic centres.

METHODS AND THEORY: MODEL STUDIES

The pulse sequence applied in DQF studies on cement pastes is shown in Fig. (**9**) of Chapter 1 (the details are described therein). A time between the first and the second 90° pulses (creation time τ_{creat}) was optimized for the maximum DQF signal. t_1 (evolution time) was kept as short as practically possible. Normally, t_1 in these experiments was a few μs. In some particular experiments, t_1 was varied (at fixed value of τ_{creat}) to monitor DQ relaxation process. In other DQF experiments, creation time was varied to study the changes of a DQF signal (at constant t_1). Full phase cycling for removal of unwanted coherences was applied according to [32].

Fig. (**3**) shows typical DQF spectra recorded for 3 different cements [5, 19, 26].

One or more pairs of anti-phase Lorentzian lines (with width w, offset d and area A) have been applied to fit the DQF spectra. For one pair this is presented according to eq. (6):

$$I(f) = \frac{2Aw}{\pi}\left(\frac{1}{w^2 - 4(f-d)^2} - \frac{1}{w^2 - 4(f-d)^2}\right) \quad \textbf{(6)}$$

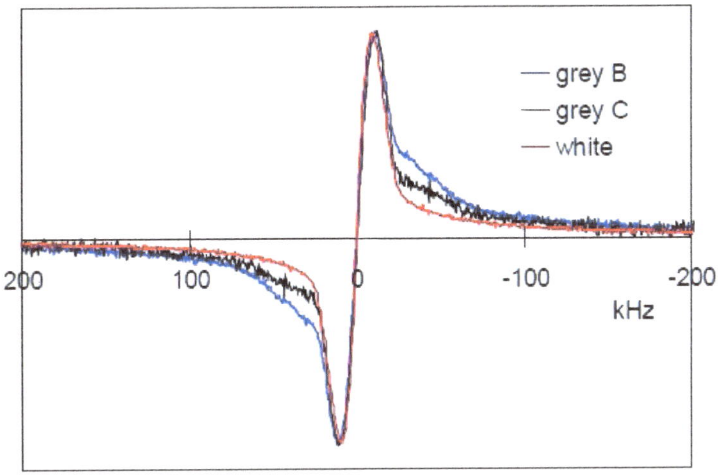

Fig. (3). The ^{1}H DQF NMR spectra for white, grey *C* and grey *B* cements after applying the anti-symmetrisation procedure $I(f) = (I_e(f) - I_e(-f))/2$.
$I_e(f)$ is experimental DQF spectrum registered as a frequency function. After phasing, the spectra were anti-symmetrised to remove any residual phasing artefact.
The content of C_4AF (Ferrite) in anhydrous cements: 0.0% (white), 7.1% (grey *C*) and 10.0% (grey *B*) [5, 25]. A creation time was optimized for maximum signal. The spectra were normalized to highlight and compare different shape lines for these cements.
The cement pastes were prepared by mixing cement powder with water (water-to-cement ratio of 0.4) according to *Nanocem* consortium procedure [4 - 7, 25]. The cylindrical moulds with the sizes of 8 mm in diameter and 20 mm deep were used for casting the samples. After the pastes were set, the samples were removed from the moulds. They were cured further under a small quantity of saturated calcium hydroxide solution at room temperature for 28 days. Prior to NMR measurement, samples were wiped by tissue paper, crushed to a coarse powder and placed in the NMR tubes. The stoppers have been used on the top of the sample to prevent an evaporation of water from the sample during NMR measurement. The sample was placed within the volume of the probe coil to ensure *RF* field homogeneity. The measurements were performed at a proton resonance frequency of 400 MHz, and at room temperature. The 90° pulse width was 8 μs. The dead time was 5 μs. Time for repetition of pulse sequence was typically 1 s. 1024 averages were acquired per spectrum. Additional studies were carried out to check that residual ^{1}H signal from mobile water was negligible compared to the DQF signals registered from cement. A signal from the probe with empty NMR tube was small. It was routinely subtracted from the total signal in each experiment [26].

Fig. (**4**) shows two examples of model DQF spectra produced by one pair of anti-phase Lorentzian lines. The DQF experimental spectra have been fitted by Lorentzian lines. The fits with Lorentzians were better than those with Gaussian function. According to [25], multiple overlapping Lorentzian functions showed relatively unstable least squares analysis. The resultant parameters were often dependent on the initial values. The DQF spectra could be fitted by similar curves, which are achieved for the different combinations of *d, w* and *A* [25]. The reason of this case was overlapping two 'up and down' Lorentzians which cancel in the middle. To avoid this instability, fitting was constrained by requiring $w = nd$ (*n* is a constant of the order of unity) [5, 25].

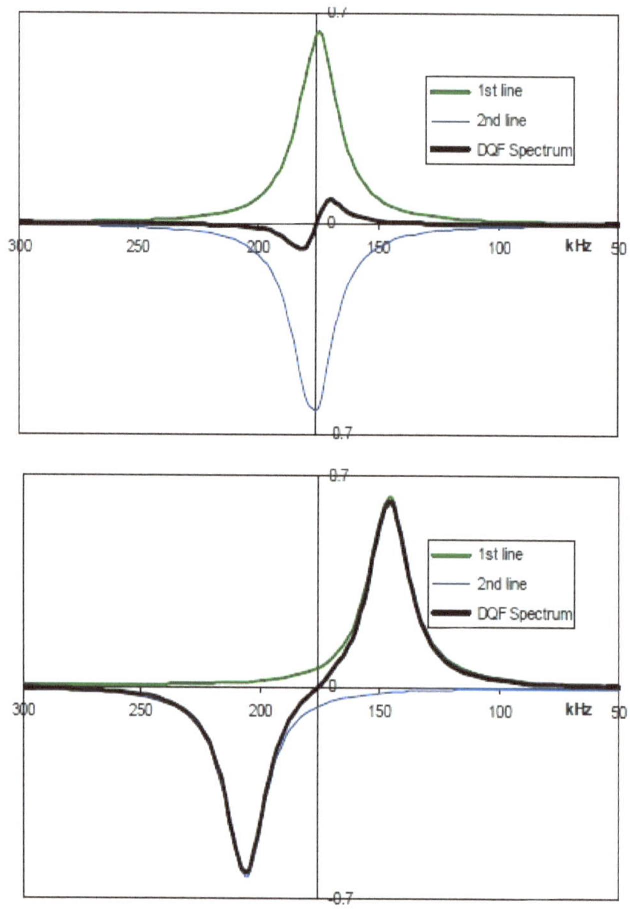

Fig. (4). The model DQF spectra produced by two anti-phase Lorentzian lines for offset *d*=1 kHz (*top*) and *d*=30 kHz (*bottom*).

For the calculation of the ^1H DQF spectrum, a model describing the water molecules, *i.e.*, proton spins I_1 and I_2, near the paramagnetic centres (Fe^{3+} ions with spin S = 5/2) has been considered [25]. The model takes into account the dipolar interactions between the two protons of a water molecule and a Fe^{3+} ion.

The dipole-dipole interaction is a function of the angles θ between the magnetic field and the vector connecting the two spins considered only. The type of the Hamiltonian governing the interaction is following [5, 33]:

$$H = \frac{C}{r^3}(1 - 3\cos^2\theta) \quad \textbf{(7)}$$

We consider a «far field» approximation in order to simplify a description of the interactions in the system of the Fe^{3+} ions and the water protons. In this case, the angles between the magnetic field direction and the vectors from the Fe^{3+} ion location towards each of the two water protons are considered as equal to θ_{IS}. This also assumes that the proton-Fe^{3+} distances are the same for both protons, *i.e.*, they are equal to r_{IS}. Fig. (**5**) shows the vector r_{IS} going from the Fe^{3+} ion location to the middle point of the distance between two protons.

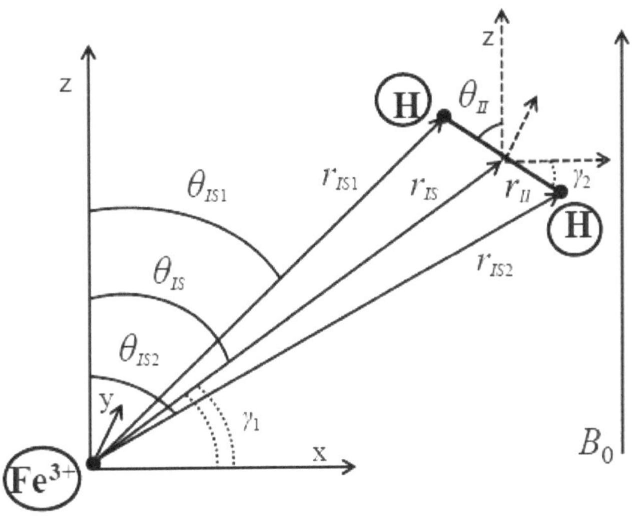

Fig. (5). A scheme of the interactions taken in the DQF model

The calculation presented in Ref. [5] obtained the FID for the system with the two considered interactions. At chosen geometry, both protons are located in the plane. A laboratory frame is defined requiring the Fe^{3+} to be at the origin and the *Z*-axis to be parallel to the magnetic field. A C-S-H sheet pore is considered to define a plane [5, 25]. The *Y*-axis in the lab frame is taken as being within the plane. The Fe-H_2O vector is in the plane. More details are in Ref. [5] showing a definition of the angles γ_1 and γ_2 that describe the relative position of the Fe^{3+} ion and protons on the plane *XY* of the coordinate frame *XYZ*. γ_1 is the angle between r_{IS} and *X*-direction, whereas γ_2 is the angle between r_{II} (vector between two protons) and *X*-direction. Because the interactions are the function of the angle between the vectors and the magnetic field only, the main interest is in calculating the angles θ_{II} (between *Z*-direction and vector connecting two protons) and θ_{IS} in the lab reference frame. To move from the gel frame to the lab frame within the geometry chosen, it is necessary to consider only a rotation around the *Y*-axis [5]. To be more precise, the interest is in the cosines of those angles, which would appear in the terms of the Hamiltonian, *i.e.*, $\cos(\gamma_1)$ and $\cos(\gamma_2)$.

The signal originating from all the water molecules on a specific plane can be calculated by integrating the FID signals for all possible angles γ_1 and γ_2. It was reasonably assumed [5, 26] that the values of γ_1 and γ_2 angles are occurring between 0 and 2π with uniform probability. In principle, an additional integral over the possible orientations of the planes can be performed also. For a randomly oriented set of vectors passing through the origin, the distribution of the angles θ_{IS} and θ_{II} between the *Z*-axis and the vectors considered is not uniform. Therefore, the additional attention (concerning that mentioned integral) should be paid to the probability of occurrence. For a geometrical visualisation, it is possible to consider the length of the circle defined by the intersection of the sphere with unitary radius and the cone defined by the angle θ with the *Z*-axis. This length is equal to $2\pi\ \sin(\theta)$ and proportional to the probability of occurrence of the direction θ [5].

The alternative approach could be considered with performing a full powder average using only two angles. The first angle is that between the magnetic field and the r_{IS} vector (θ_{IS}), whereas the second is the angle between the magnetic field and the r_{II} vector (θ_{II}). Ref. [5] assumed cylindrical symmetry for both angles. Then, a probability of occurring was proportional to the *sine* of the angle used.

In the calculation of magnetic interaction, following basis has been used [25]:

$$|1\rangle = |\alpha\alpha, m_z\rangle;\quad |2\rangle = \left|\frac{\alpha\beta+\beta\alpha}{\sqrt{2}}, m_z\right\rangle;\quad |3\rangle = |\beta\beta, m_z\rangle;\quad |4\rangle = \left|\frac{\alpha\beta-\beta\alpha}{\sqrt{2}}, m_z\right\rangle \tag{8}$$

As the spin of the Fe^{3+} ion is equal to 5/2, m_s can accept the values -5/2, -3/2, -1/2, 1/2, 3/2, 5/2. α and β refers to the state of proton «spin up» and «spin down», respectively.

A dimension of the basis state is 24. Therefore, the matrices used as evolution, density, and rotation matrices are 24×24. The Hamiltonians used for proton-proton interaction of like spins are truncated. For the proton-electron interaction, the Hamiltonians are also truncated in the weak coupling approximation (for unlike spins). Therefore, the states for different m_s do not interact. The calculation has been performed for each m_s value separately. After the calculations for all m_s values, the results have been summed [5]. This is equivalent to limit the work on the 4×4 block matrices on the diagonal of the full matrices (Fig. **6**). The fourth state in eq. (8) does not participate in NMR measurements because it does not carry magnetization [5]. Thus, the representation could be simplified using six 3×3 matrices, which result in magnetization, and six 1×1 matrices, which do not give that. Therefore, for last case, it was not necessary to perform the calculation [5].

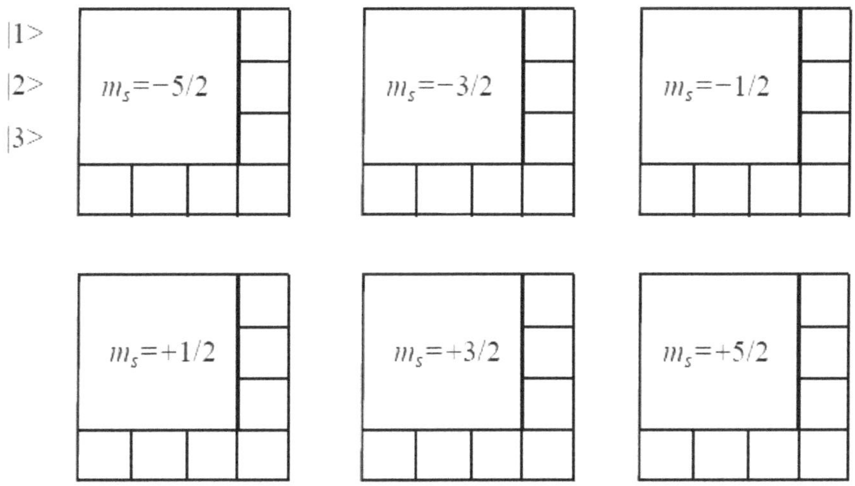

Fig. (6). Block matrices 4×4 with different m_s values: -5/2, -3/2, -1/2, +1/2, +3/2, +5/2. These 6 matrices are placed on the diagonal of the 24×24 matrices describing the system simplified with the assumed approximations. The rest 1×1 boxes outside the diagonal of the 24×24 matrices are empty. The empty boxes do not take participation in the NMR experiment [5].

The truncated form of the Hamiltonian used for the proton-proton (*I-I*) interaction is given by eq. (9) [5, 33]:

$$H_{II} = \frac{\mu_0}{4\pi}\frac{\gamma_I^2\hbar^2}{r_{II}^3}\left(\frac{1-3\cos^2\theta_{II}}{2}\right)(3I_{1z}I_{2z} - I_1 I_2) = C_{II}(3I_{1z}I_{2z} - I_1 I_2) \quad \textbf{(9)}$$

For the *I-S* interaction between Fe^{3+} ion and proton (in the weak coupling approximation), the Hamiltonian is used according to eq. (10):

$$H_{IS} = \frac{\mu_0}{4\pi}\frac{\gamma_I\gamma_S\hbar^2}{r_{IS}^3}\left(\frac{1-3\cos^2\theta_{IS}}{2}\right)(I_z S_z) = C_{IS}(I_z S_z)$$
$$\left(\frac{\mu_0}{4\pi}\frac{\gamma_I\gamma_S\hbar^2}{r_{IS}^3} \ll |\omega_I - \omega_S|\right) \quad \textbf{(10)}$$

The vectors r_{II}, r_{IS}, and the angles θ_{II}, θ_{IS} presented in the eq. (9) and eq. (10) were defined above in the text. *I* and *S* are the spin angular momentum operators. The remaining constants are μ_0, the magnetic permeability of free space, $\hbar$, Planck's constant divided by 2π, and $\gamma_{I,S}$, the magnetogyric ratio of the nuclear and electron spins, respectively. In these approximations, the dipole-dipole Hamiltonians are diagonal and the calculation is quite simple [25].

The assumption of some values for the geometric parameters used in the calculations [5, 25] (namely, $r_{IS1} = r_{IS2} = r_{IS}$, and $\theta_{IS1} = \theta_{IS2} = \theta_{IS}$) is quite good because the pair of protons considered is far enough from paramagnetic centre, and $r_{IS} >> r_{II}$. The exception is only for the case of very small distance between the water molecule and the paramagnetic centre. Such a very close approach of the water molecule to the Fe^{3+} ion corresponds to the widest regions of the spectral line [25]. The fraction of protons with the closest distance to the paramagnetic ion in the calculation of the spectrum according to powder averaging was small indeed. Thus, this is not a main limitation to the accuracy of the calculations. More details in theory and calculations are in Refs. [5].

The full width varied as $1/r_{IS}^3$ for the case of small distances. At large distances, the spectrum reduced to that of an isolated water molecule. A numerically integrated spectrum for water in a powder average of planar pores with a cut-off radius r_c = 24 Å was obtained for comparison [5].

At large r_{IS}, the spectrum is reminiscent of an anti-symmetric version of the symmetric two-line spectrum that is observed in single-quantum NMR experiments of oriented, rigid dipolar-coupled spin-1/2 pairs, such as in a single crystal of gypsum. The magnetic dipolar field of one nucleus shifts the resonance of its partner by an amount dependent on the distance between the two nuclei and the orientation of the inter-nuclear vector relative to the applied magnetic field. The powdered patterns for spectra are available rather than those for single crystals. The distance between different pairs of nuclei is constant. However, the relative orientations are not constant.

The observed spectrum comprises a powder average over all orientations. Since the probability of the angle between the inter-nuclear vector and the field (θ_{II}) varies as $\sin(\theta_{II})$, some splitting is far more likely than others. As a result, the powder spectrum exhibits a complex shape dominated by two strong lines or singularities corresponding to $\theta_{II} = \pi/2$. For the case of the single-quantum experiment, the shape is the well-known Pake doublet [34, 35].

DOUBLE-QUANTUM-FILTERED (DQF) SIGNALS: DEPENDENCE ON SAMPLE CURING AND ON CREATION TIME. PAKE DOUBLET

$Ca(OH)_2$ is a product which is developing during cement hydration. Therefore, for understanding the DQF spectra of cements, the DQF spectrum of calcium hydroxide has been also calculated [25]. For this case, a proton-proton distance from the characteristic one for water (1.6 Å) has been changed to interproton distance for $Ca(OH)_2$ (2.186 Å). The value of $r_{IS} = \infty$ has been used in the programme of calculation to eliminate an effect of paramagnetic impurities. Fig. (**7**) shows the powder averaged DQF spectra for $Ca(OH)_2$ obtained for different τ_{creat} values.

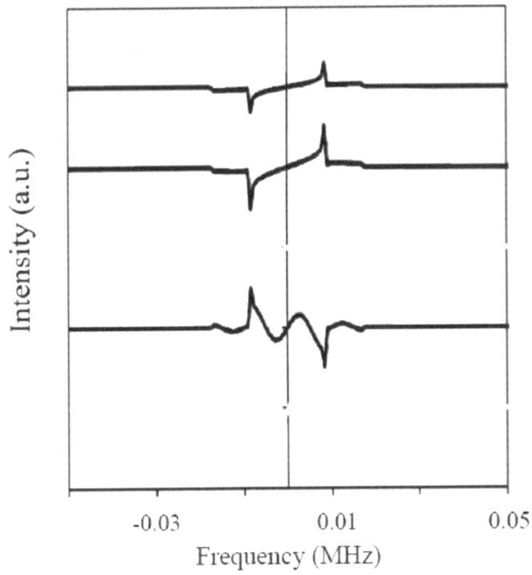

Fig. (7). Series of the ^{1}H DQF spectra calculated for different creation times: 10 μs (top); 20 μs (centre) and 100 μs (bottom). The distance $r_{II} = 2.186 \cdot 10^{-10}$ m (data for $Ca(OH)_2$ according to Ref. [36]). 2500 steps have been calculated for powder average with angles of θ_{II} between 0 and π. Evolution time is 10 μs for all calculations with different creation times. $r_{IS} = \infty$, *i.e.*, the protons are far from the Fe^{3+} ion.

It is seen from Fig. (**7**) that the wiggles appear with increasing a creation time, *e.g.*, at $\tau_{creat} \geq 100$ μs. The creation time $(\tau_{creat})_{max}$ is the value to obtain maximum signal intensity for the $Ca(OH)_2$ simulated spectra. It has been found that for $Ca(OH)_2$ the $(\tau_{creat})_{max} = 30$ μs. Fig. (**8**) shows the effect of the different proton-proton distances on the DQF spectra. The splitting between positive and negative peaks can be compared when the proton-proton distance is changed. These simulated spectra (Fig. **8**) were obtained also without the effect of the Fe^{3+} ions. An influence of the Fe^{3+}, *i.e.*, a consideration of various Fe-water distances (r_{IS}) in simulations results in much more complex DQF spectra. Fig. (**9**) shows this effect for one particular set of angles ($\theta_{II} = \theta_{IS} = \pi/2$).

As the Fe^{3+}-water distance tends to 0, the spreading of frequencies becomes big. In order to consider all factors, the powder average must be performed also for the Fe^{3+}-water angle. Then, there will be an additional variable which is the Fe^{3+}-water distance [5, 25]. When the Fe^{3+}-water distance increases, the shape of the spectra converges to the one for long distance as the one presented in Fig. (**7**) for long creation time.

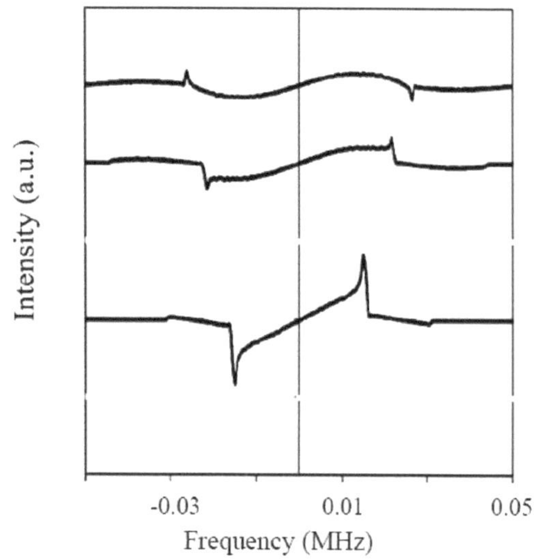

Fig. (8). The ^{1}H DQF spectra calculated for different proton-proton distances: $r_{II} = 1$ Å (top), 2 Å (centre), and 4 Å (bottom). The creation time is 20 μs. For the θ_{II} powder average, 2500 steps have been used. $r_{IS} = \infty$, *i.e.*, the protons are far from the Fe^{3+} ion, and the effect of paramagnetic impurity has not been considered in calculations.

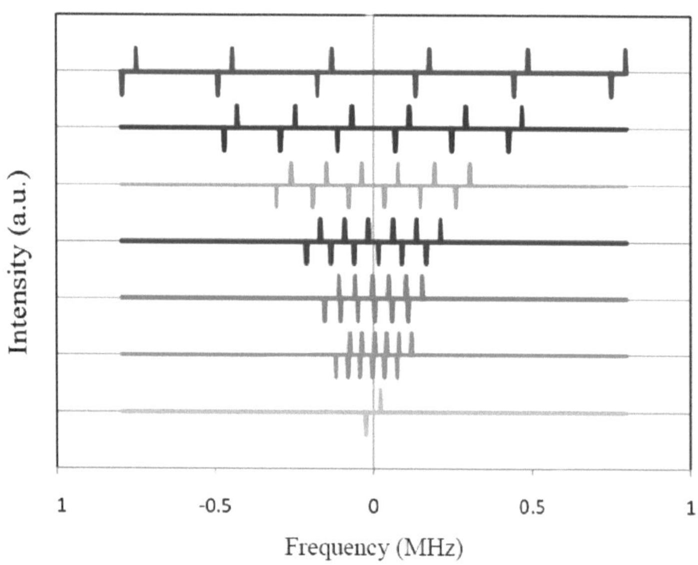

Fig. (9). The DQF spectra calculated for different distances r_{IS} between water molecule and Fe^{3+} ion. r_{IS} increases from top to bottom: 1.6 Å, 4.8 Å, 8 Å, 11.2 Å, 14.4 Å, 17.6 Å, ∞. The creation time is 55 μs. The proton-proton distance $r_{II} = 1.6$ Å. In the last plot (bottom), the water molecules are far from the Fe^{3+} ion The simulation has been performed for the angles: $\theta_{II} = \theta_{IS} = \pi/2$.

The obtained DQF spectrum is the superposition of DQF spectra for different Fe^{3+}-water distance, indeed. The simple geometrical model that has been used to calculate the averaged spectra contains a Fe^{3+} ion located in the centre of a sphere and several water molecules uniformly distributed in half of the sphere [5, 19, 26].

Fig. (**10**) shows the ^{1}H NMR DQF spectra of white Portland cement (*left*) and also the changes of the intensity of these DQF spectra with curing time (*right*). An increase of the ^{1}H NMR DQF intensity during curing was in line with the SQ study, which measured spin-spin relaxation time and showed the growth of the population of the short component (T_2 is about 40-50 μs or less). In addition to the SQ (T_2) data on cement hydration, similar SQ experiments have been also carried out in other publications [5, 6, 8, 20, 11, 37]. When the growth of the amplitude of the short T_2 component with curing time occurred, the loss of the intensity of the components with long spin-spin relaxation time was also observed. Normally, the long T_2 components could be measured quite easily. Fig. (**11**) shows the ^{1}H SQ NMR spectra of white cement at different curing times. The spectra mirror behaviour of the long T_2 component and the intensity of the long T_2 component as a function of hydration time. The SQ study (Fig. **11**) was carried out in parallel to the DQF studies of hydrating on white cement [25, 26].

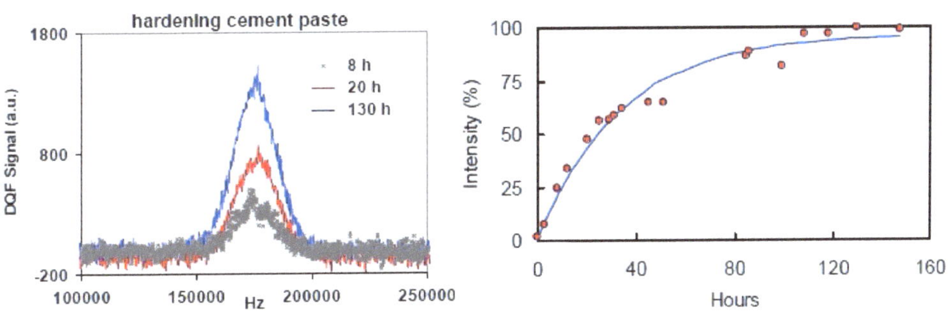

Fig. (10). The ^{1}H DQF spectra (*left*) ("*in phase*" mode), and the DQF signal intensities (*right*) for the white cement with early ages (up to 6 days) *vs* curing time. The samples were cured in the NMR tubes after mixing cement powder and water (w/c=0.4). Between measurements, the samples were placed under a small reservoir of the calcium hydroxide solution. Further, hydration reaction was monitored measuring the NMR signal of the same sample. An increase of the DQF signal amplitude with curing time confirms a creation of anisotropic structure. All measurements were carried out at room temperature using a 400 MHz NMR spectrometer.

These results were consistent with the study of mobile water in hardening cement paste [8, 25]. The mobile water was becoming confined and chemically bound in the hydration products of the cement [26, 5, 38]. Thus, this confirmed that DQF NMR [19, 25] could measure the confined and bound water fraction.

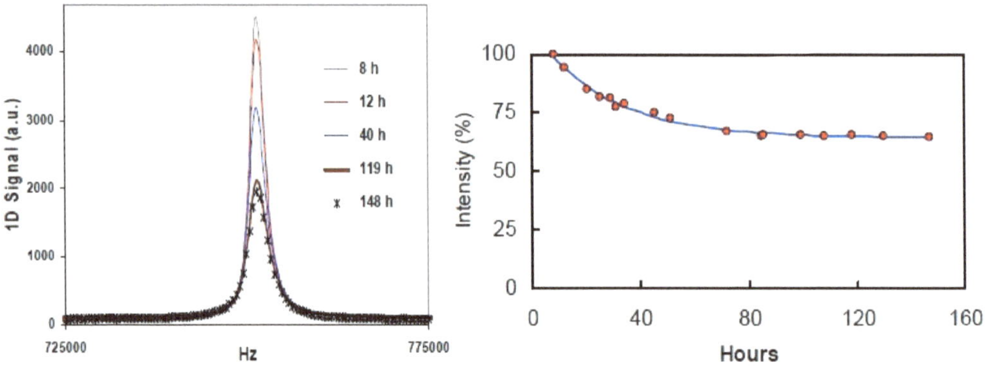

Fig. (11). The SQ spectra (*left*) and the signal intensities of this mobile signal (*right*) for white cement *vs* curing time. The decrease of the amplitude of normal SQ signal with curing time informs the loss of water mobility with curing. The NMR experiments were made at proton frequency of 400 MHz, and at room temperature.

Other data presented in Ref. [25] showed the dependence of the total DQF signal intensity on the τ_{creat} defined as the integral area of the half spectrum (Fig. **4**). For all the three cements studied (Fig. **3**), and for the $Ca(OH)_2$, the samples had a single maximum in the DQF signal intensity occurring between 33 and 40 μs. For instance, Fig. (**12**) shows the series of the ^{1}H DQF spectra for grey cement C at the various creation times as a 3D plot. The maximum intensity of the DQF spectrum in this series was registered at τ_{creat} = 38-39 μs.

Fig. (12). The ^{1}H DQF NMR spectra of hydrated and powdered grey cement C specimen at varying creation times from 32 μs to 2 ms. The frequency is 400 MHz. T=25°C.

Fig. (**13**) shows ^{1}H SQ NMR spectra for hydrated white cement and pure $Ca(OH)_2$ powder. At presented drying step of cement sample, a broad component is observable. This broad component is about several tens of kHz. This indicates that it originates from the protons of a solid matrix. The intensity of the narrow liquid like line is small. However, proton SQ NMR spectrum of cement has the shape that is still determined by the presence of the mobile water protons.

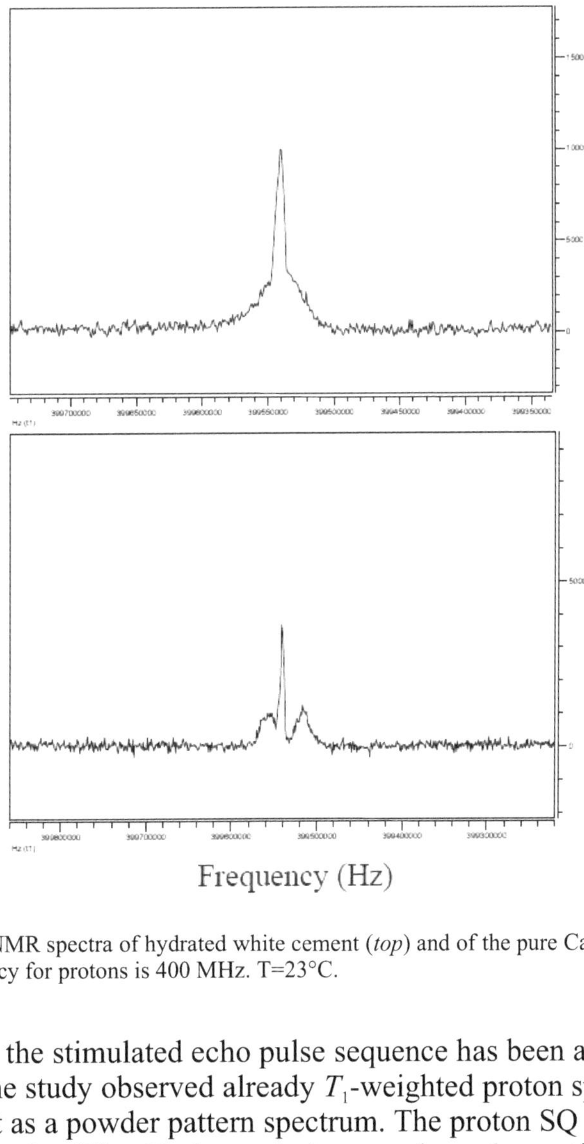

Fig. (13). The ^1H SQ NMR spectra of hydrated white cement (*top*) and of the pure $Ca(OH)_2$ powder (*bottom*). The resonance frequency for protons is 400 MHz. T=23°C.

In the work [12], the stimulated echo pulse sequence has been applied in studying white cement. The study observed already T_1-weighted proton spectrum recording broad component as a powder pattern spectrum. The proton SQ NMR spectrum of pure $Ca(OH)_2$ powder (Fig. **13**, *bottom*) shows a shape (excepting small signal in the centre from the protons of residual surface water) which is similar to a crystalline water doublet structure, *e.g.*, like to that in a gypsum powder $CaSO_4{\cdot}2H_2O$ [12, 34]. In the solids, the proton SQ NMR spectrum is determined by magnetic dipolar interactions. Therefore, it is dependent upon the location of protons within solid sample. There is a random orientation of the nuclei in the sample. As a result, the proton SQ NMR spectrum of solids is a powder pattern spectrum [34]. Because of the difficulties to study the solid protons, an additional NMR method is seemed to be useful to clarify water state in hydrated cement and calcium hydroxide with more details [8, 19, 39 - 41]. In particular, Ref. [8] presented comprehensive exploration of the population of the fast and slow relaxing components of spin-spin relaxation time of protons in $Ca(OH)_2$ and cement pastes to detail solid and mobile components of water and microstructure of hydrated cements.

The solid echo pulse sequence consists of the two RF pulses, *i.e.*, $(\pi/2)_x$-τ-$(\pi/2)_y$ acquisition (where τ is a short delay interval) [42, 43]. Ref. [8] gives more details in description of the solid echo experiment on cement samples. After the second pulse of the 20 MHz Benchtop NMR Instrument (dead time is ~ 9-11 μs), the free induction decay and solid echo signal are recorded. At usual registration of FID, the signal would decay rapidly due to rigid dipole-dipole interactions. If to use the $(\pi/2)_x$-τ-$(\pi/2)_y$ experiment, the signal from the solid protons results in an echo [8, 42, 44]. As it was showed by Powles and Strange [44], for rigid coupled spin hydrogen pairs such as in solid $Ca(OH)_2$, the solid echo pulse sequence refocuses signal of magnetisation. The echo has roughly Gaussian shape with the centre on a time τ after the 2nd $(\pi/2)_y$ pulse. The composite signal with mobile and solid components can be analysed by fitting the data to the equation presented in the Refs. [5, 8, 42].

Fig. (**14**) shows a solid echo signal measured from dried and powdered $Ca(OH)_2$ at τ=20 μs. The echo decays rapidly. After that there is a small negative oscillation known as a Pake doublet [25, 35]. Pake doublets show the features of systems comprising rigid pairs of dipolar coupled spin 1/2 nuclei, such as hydrogen protons in calcium hydroxide [34, 35, 38]. Due to incomplete refocusing of the echo, it is necessary to do a back extrapolation of the experimental dependence of solid echo intensity *vs* (τ) towards τ = 0 to get the right value of the intensity of solid component [8, 42] (Fig. **15**).

The nearest neighbours have the dominant effect on the dipolar interactions, *i.e.*, the model describing the SQ NMR spectrum should be determined by the nearest neighbour interactions, broadened by additional dipole-dipole interactions from far nuclei. Thus, for two spins with dipolar interaction at a distance *r*, the powder SQ NMR spectrum has the Pake doublet line shape [5, 12, 33 - 35].

According to [38], water in hardened Portland cement can be characterised by three types: chemically bound, physically bound and porous trapped water. Authors heated hardened cement at 105°C and discovered the loss of the physically bound water while the chemically bound water was in a stable form

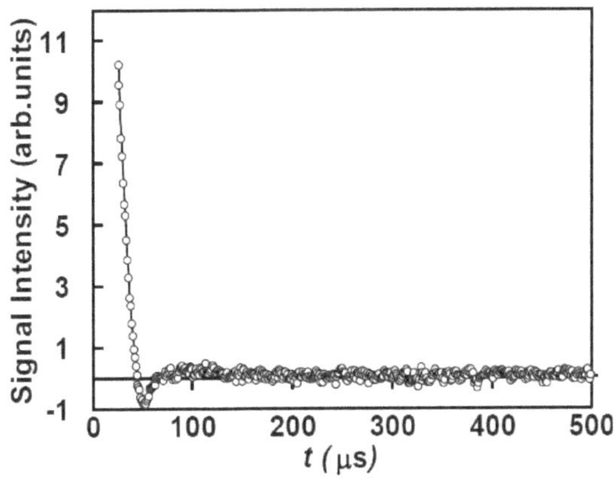

Fig. (14). The solid echo signal in the experiment $(\pi/2)_x$-τ-$(\pi/2)_y$ for dried $Ca(OH)_2$ [8]. Time delay τ=20 μs. The oscillation is typical in time domain, whereas a Pake doublet is registered in frequency domain. Resonance frequency for protons is 20 MHz. The temperature of measurement is 25℃.

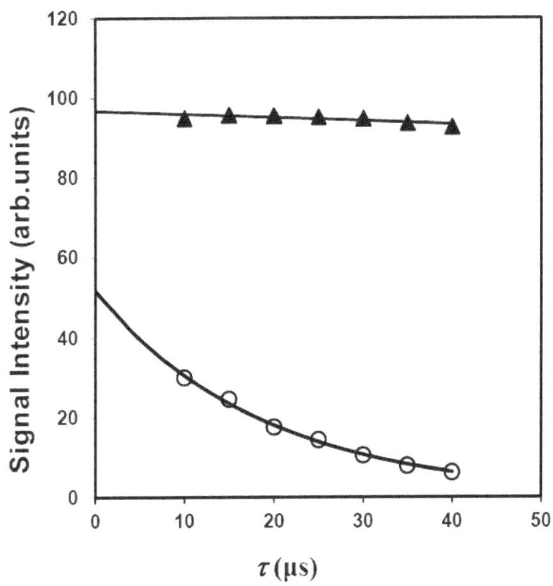

Fig. (15). The intensities of the solid echo (circles) and mobile (filled triangles) signals for the paste of white cement A *vs* τ (a delay between RF pulses in solid echo experiment). The sample was mixed with a w/c ratio of 0.4 and hydrated (28 days) at a room temperature. For NMR measurement, samples was wiped by tissue paper, crushed to a coarse powder and placed in the NMR tube. The stopper has been placed in the NMR tube on the top of the sample to prevent an evaporation of water from the sample during measurement. The sample was placed within the volume of the probe coil to ensure RF field homogeneity. The resonance frequency for protons is 20 MHz. The temperature of measurement is 25℃.
All measured intensities were normalised per unit mass. The data were fitted by exponential decays (the solid lines). The extrapolation of the exponential fit for the solid echo intensity to delay $\tau = 0$ resulted in right value of population of solid component.

[38]. They observed a Pake doublet in the ^{1}H NMR spectra explaining the results by the oscillation of the water molecules and hindering molecular motions because of the entrapment of water in the pores of cement. The water used in the formation of $Ca(OH)_2$ was considered as «chemically bound» water [38]. $Ca(OH)_2$ is formed by the reaction of water with calcium oxide, and the molecular water motion in this reaction is the slowest among all water types mentioned above.

The proton powder spectrum of pure $Ca(OH)_2$ dried at 105 °C showed two humps [38]. The one to the right of the –OH narrowing is responsible for the anisotropic chemical shift along *X*- and *Y*-axes. Based on crystal symmetry of $Ca(OH)_2$, the authors of Ref. [38] consider that the axes of *X* and *Y* are equivalent. The peak of the spectrum which is to the left of the –OH narrowing is connected to the anisotropic chemical shift along the *Z*-axis. These authors considered that spin density in the *X*-*Y* plane exceeded the spin density along the *Z*-axis. This could be the reason that the right peak is more intensive than the left one [38].

It is possible to apply the model of water molecules which are in random movement across the surface of the planar pore [4, 25]. Moreover, the water molecules in the pores interact with the paramagnetic Fe^{3+} centres. Therefore, the model has been developed further to be applied to ^{1}H DQF NMR data [25, 26, 19, 39 - 41]. In particular, the angle between the normal to the pore surface and the inter-proton vector was considered as constant [25, 40].

The orientations of the pore are obeyed to a powder average pattern. The pore edges can bind the water movement [26]. In the case that Fe^{3+} amount is quite high or when the edges of the pore restrict the water walk, then the resultant spectrum of a powder pattern can be considered as a weighted sum over all distances. The closest approach of the water molecule to the Fe^{3+} ion is considered as r_{min} = 30 Å [4, 5], and a maximum cut-off radius is r_c. It is equal to the radius of the pore. Ref. [25] reported the numerically integrated powder spectrum for r_c = 24 Å. Further details of the model and calculations are in Refs. [5, 25].

DQF LINE SHAPE

The analysis of the DQF spectrum line for $Ca(OH)_2$ was done in [26, 19, 5, 40, 41], when it was recorded under identical conditions to the white cement. A fit of the DQF data for calcium hydroxide according to eq. (6) discovered that the ^{1}H DQF NMR spectrum contains a pair of anti-phase signals described by Lorentzian functions. Calcium hydroxide has a crystalline structure and fitting the DQF spectrum of $Ca(OH)_2$ results in an asymmetric version of the SQ Pake doublet [35] presenting separated negative minimum and positive maximum [25, 40]. Ref. [25] compared also the measured DQF NMR spectrum for $Ca(OH)_2$ powder with the theoretically calculated DQF spectrum without broadening. An inter-hydrogen distance of 1.75 Å has been used in the calculation. Some variations in the inter-hydrogen distance (in the range of 2.02 to 2.18 Å) in $Ca(OH)_2$ were reported in literature [36, 45]. This distance as determined by NMR measurements differs from that obtained in neutron scattering experiments [45]. There is a strong broadening effect in the NMR data [25]. An additional factor affecting NMR spectra is a spectrometer dead time. SQ Pake doublet showed similar broadening effect. The numerical simulations of the broadening Pake doublets have been performed in the work [46]. Taking these results into account, the broadening may be estimated. When broadening is known, a correction can be made for spectrometer dead time [46]. If to apply the data [46] to the SQ Pake doublet spectrum [5, 26, 34, 40, 41], then it is possible to find that both effects are strong. However, the one effect acts in opposition to another. More details are in the Ref. [25].

The work [25] compared the typical ^{1}H DQF NMR spectrum of white cement (aged of 28 days, measured at τ_{creat} = 37 μs) with the DQF spectrum of powdered $Ca(OH)_2$ recorded at experimental conditions identical to those at registration of white cement spectrum. Fitting the DQF data of white cement has been probed similarly to $Ca(OH)_2$ as a pair of anti-phase Lorentzian lines. However, an analysis of the DQF spectra showed that the spectral width in the wings of cement spectrum is relatively lager compared to calcium hydroxide DQF line. The line shape and the fit were discussed in details in Refs. [5, 25].

As it was found in Ref. [25], the more narrow component in the DQF spectrum for Portland white cement is slightly narrower than that for pure $Ca(OH)_2$. This is in line with the results of the work [12] for the Pake doublet. There is a slight difference in lattice structure between the $Ca(OH)_2$ in pure powder and that in cement. The lattice spacing in cement is a bit modified. This is because it occurs only as nanocrystals for which the surface energy plays a more significant role. The results obtained [5, 19, 25, 41] suggested that the DQF spectrum of white cement contains a linear combination of two components. The first component is a narrow anti-phase Lorentzian doublet which is responsible for the crystalline structure of $Ca(OH)_2$. The second component is a broader anti-phase Lorentzian doublet which is responsible for the water confined in the planar pores of C-S-H gel.

The DQF spectra for grey C and grey B cements were compared with those for white cement [25]. In the anhydrous materials of grey C and grey B cements the Ferrite concentration increases from grey C to grey B. If to compare Fe content in white and grey cements, then one should emphasize that white cement practically has no Fe^{3+} impurity. The spectra of both grey cements were recorded under identical conditions to those for white cement and $Ca(OH)_2$. The data for grey cements C and B show clearly the additional broad line (shoulder) in the DQF spectra. The intensity of this line (shoulder) could be considered as indicator of Fe content. The increased intensity of this broad line was recorded for grey cement B which has an increased content of mineral phases AFm and AFt [47].

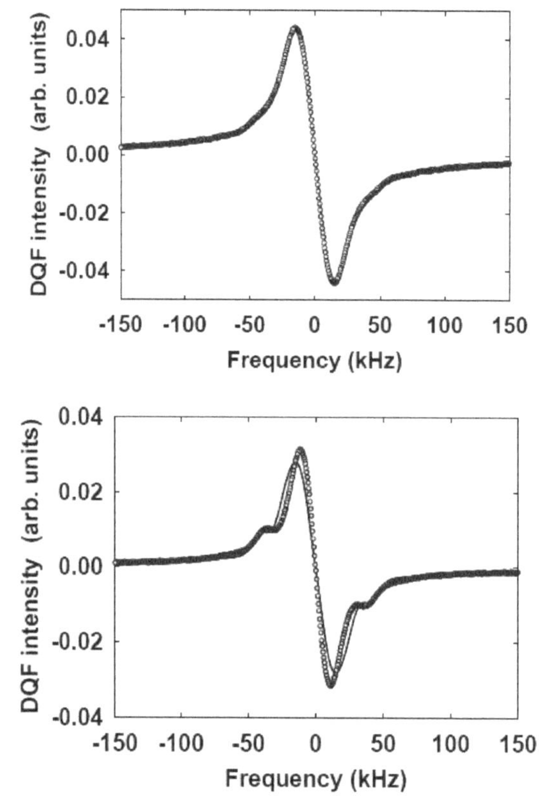

Fig. (16). The measured ^1H DQF NMR spectra (400 MHz) of hydrated grey cement C (*top*) and grey cement B (*bottom*) at room temperature. The fits of the data are presented as a linear combination of the 1st Lorentzian anti-phase pair (with parameters of $Ca(OH)_2$ spectrum), 2nd Lorentzian anti-phase pair (the spectrum of water in C-S-H), and 3rd Lorentzian anti-phase pair (the spectrum of water in the phases with Fe-impurities). The DQF spectra for grey cements C and B were registered with identical experimental parameters.

Fig. (**16**) shows the ^1H DQF NMR spectra for grey cements C and B. The fit of the data for grey cements based on the eq.(6) was done according to the linear combination of spectral lines presented for DQF spectrum of white cement. In addition to $Ca(OH)_2$ spectral part (1st Lorentzian anti-phase pair) and C-S-H line (2nd Lorentzian anti-phase pair), the DQF lines in the grey cements contained third Lorentzian pair of anti-phase doublet responsible for the iron-enriched phases.

DQF AND DRYING THE SAMPLES: EXCHANGEABLE WATER

Fig. (**17**) shows the dependences of a DQF signal intensity *vs* creation time for hydrated white cement in the experiments upon drying. The top curve (where the DQF signal intensities have maximal amplitudes in comparison to the intensities of other build-up DQF curves) is for the cement sample before drying [19, 40].

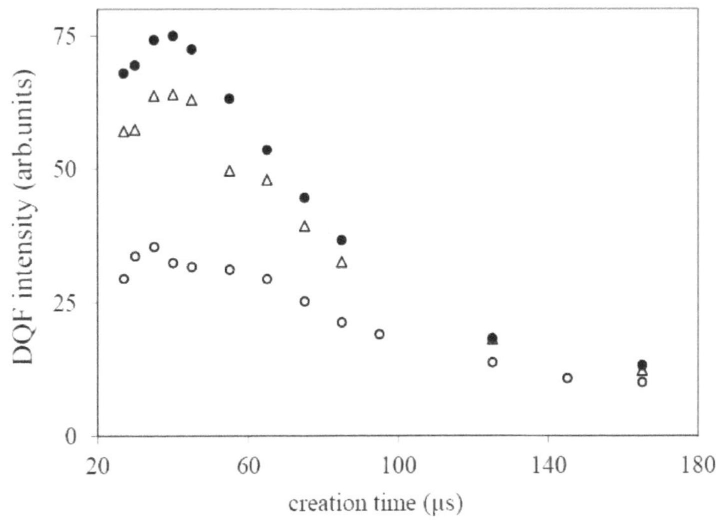

Fig. (17). The total intensity of ^{1}H DQF NMR spectra of white cement *vs* creation time for different steps of drying: hydrated cement (aged 28 days) before the start of drying (closed circles), after heating at 300 ℃ (open triangles) and after additional heating at 350 ℃ (open circles). The intensity was defined as the integral area of the half (positive part) spectrum. The NMR measurements were carried out at 23 ℃.
For the progressive drying study, the samples were weighed before and after drying in an oven. Drying was initially for a few minutes just above ambient temperature, gradually increasing to longer times at temperatures of up to 200 ℃ (vacuum oven). An oven without a vacuum was used for higher temperatures: 300, 350 and 400 ℃.

The DQF signal is responsible for the anisotropic bound water, and it is decreasing with drying (heating). In particular, the strong decrease in the DQF intensities is observed in the range of creation times around $(\tau_{creat})_{max}$, *i.e.*, in the area of the maximum of DQF signal. The experiment with progressive heating showed how anisotropic water leaves from cement sample (Fig. **17**).

Fig. (**18**) shows the DQF signal intensities in the samples of white and grey cements *vs* relative sample mass during drying. These DQF data for both cements are qualitatively comparable with the data from solid echo experiment on white cement, in particular, a population of the fast relaxing (solid) T_2 component had a similar dependence on the sample mass during drying cement [5, 8]. The intensity of the solid component during drying increased a bit at early stages of drying and was decreasing at relative sample mass less 0.84. When water leaves at relatively small sample mass, the DQF signal decreased with an accelerating rate.

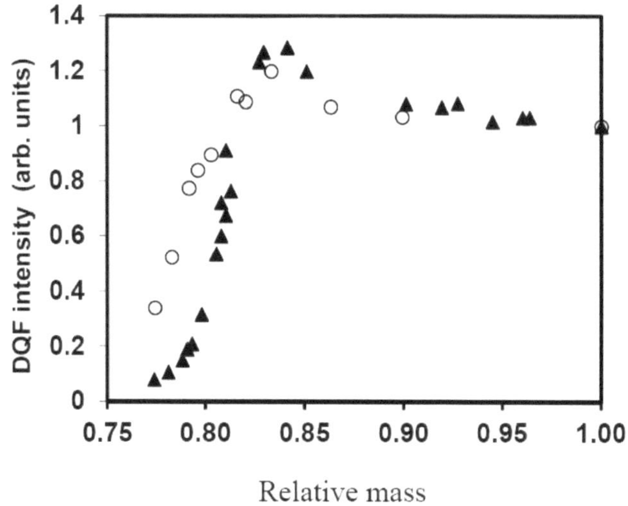

Fig. (18). The DQF spectra intensities as the function of relative sample mass for the progressively dried samples of white cement A (closed triangles) and grey cement B (open circles) measured at 400 MHz and T=23 ℃. The DQF intensities of each set were normalized per original DQF intensity of the sample at the start of drying. This shows obvious initial increase of the DQF signal as water is lost. These behaviours for both cements are in line with the data of solid echo experiments [8] (measured at 20 MHz) which mirrored the solid T_2 component dependence in white hydrated cement at early increase in signal and a sharp decrease of that at later drying stages.

Ref. [25] showed that the DQF spectrum in white cement is fitted to two components attributed to water in C–S–H gel pores and to the protons of $Ca(OH)_2$. It was shown [19, 25, 40] that the component assigned to the solid $Ca(OH)_2$ first increases before decreasing, while that associated with the C–S–H protons decreases monotonically. This increased DQF intensity for solid $Ca(OH)_2$ can be explained as follows. When the evaporated water leaves, the adsorbed layer remains. The adsorbed water has DQF features which are closely comparable to those of $Ca(OH)_2$. Thus, an increase of the population of solid component (registered as increased DQF intensity) is due to the participation of the adsorbed water in the early steps of drying process.

After removing mobile water, the water of the adsorbed layer is also lost (at higher temperatures). Thus, the crystalline products break down. This leads to the loss of the DQF signal. Separate attention should be paid to the analysis of the results obtained at drying temperatures in the range of less than 200 °C and above 200 °C. At the temperatures <200 °C, an oven with a vacuum pump was available for experiments with drying cement. Therefore, carbonation did not seem to be a problem for the DQF study of cement drying using this oven. For drying above 200°C, the oven was without vacuum pump. Therefore, for this temperature regime, carbonation can give a serious issue. In total, these results obtained are in line with those of the study published in Refs. [8, 26, 40, 41]. The publication [8] studied the SQ intensity during progressive drying white cement paste. Those results presented a decomposition of SQ intensity into slow and fast relaxing T_2 components using the experiments with solid echo and FID. The intensity of a long component (with T_2 > 60 μs) decreases faster than the total, while the population of the short component (the T_2 is less than ~ 40 μs) increased initially as water was removed.

In order to clarify, which kind of water in cement samples is responsible for the measurable DQF signal, the experiments with H_2O/D_2O exchange have been carried out. After hydration of cements, the samples were crushed into small pieces, which then were placed in the vial with D_2O for 10-15 min. After this wash in D_2O, surface water on cement pieces was removed by tissue paper. NMR measurement was looking for ^1H DQF signal in washed cement sample. After measurement, a next wash in D_2O has been carried out. Totally, 7-9 washes were carried out (until residual DQF signal became constant). Fig. (**19**) shows a DQF signal intensity *vs* creation time for hydrated grey cement and for the same sample after several washes in D_2O. The comparison of these two curves shows that H_2O/D_2O exchange can decrease the initial DQF signal of protons in measurable sample in 5-6 times. Fig. (**20**) shows the proton DQF NMR spectra for hydrated white cement sample in H_2O/D_2O exchange experiments at creation time resulted in maximum DQF signal. After last wash in D_2O, the integral intensity of a DQF spectrum was ~18% of initial DQF signal in the sample before start of the D_2O/H_2O exchange experiment. For different cement samples, D_2O/H_2O exchange experiments resulted in residual DQF signal in the range of 16-22%. In these experiments, the DQF signal of protons has identified anisotropic water which does exchange as evidenced by D_2O. In addition to this, further, anisotropic water is giving a residual DQF signal which does not exchange. It supposes a possibility to differ chemically bound water that did not exchange. The protons, which are associated with non-vanished DQF signal in exchange experiment, probably, stem from OH groups of $Ca(OH)_2$.

Additional experiments which combined drying (heating) approach and washes in D_2O with the aim to remove anisotropic water and differ this from chemically bound water resulted in the data showed in Fig. (**21**). This is a comparison of the DQF data for two white cement samples: one sample started the cycle of D_2O washes when it still had relatively intensive DQF signal after heating at 190 °C, whereas second sample was additionally heating in oven at 390 °C. That additional heating at high temperature resulted in substantial decrease of the DQF signal at the start of D_2O washes. However, the final D_2O washes resulted in comparable DQF signals for both cement samples. The results testified that there is a residual DQF signal (after all procedures with cement samples) of circa 18% of the initial maximum that cannot be removed by drying at temperatures up to ~400 °C or by multiple exchange washes with D_2O.

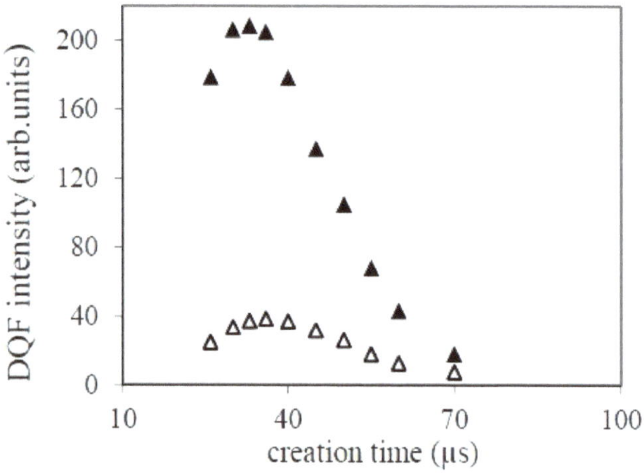

Fig. (19). The creation time dependences of the DQF intensity for grey C cement before H_2O/D_2O exchange (closed triangles) and after 5 washes in D_2O (open triangles). The NMR measurements (400 MHz) were carried out at 23 ℃.

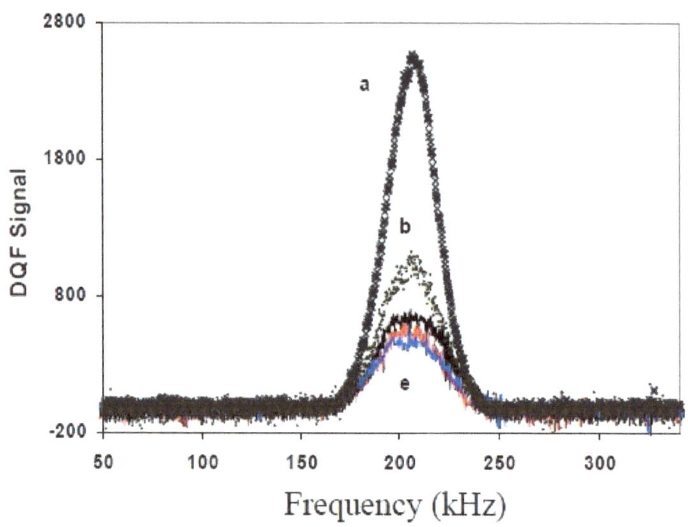

Fig. (20). ^{1}H DQF NMR spectra ("*in phase*" mode) for hydrated white cement in D_2O/H_2O exchange experiment for several D_2O washes; the most distinct steps are marked: **a** – original cement sample before the start of D_2O/H_2O exchange; **b** –after 1 wash in D_2O; **e** – after 7 washes in D_2O. The NMR measurements (400 MHz) were carried out at 23 ℃.

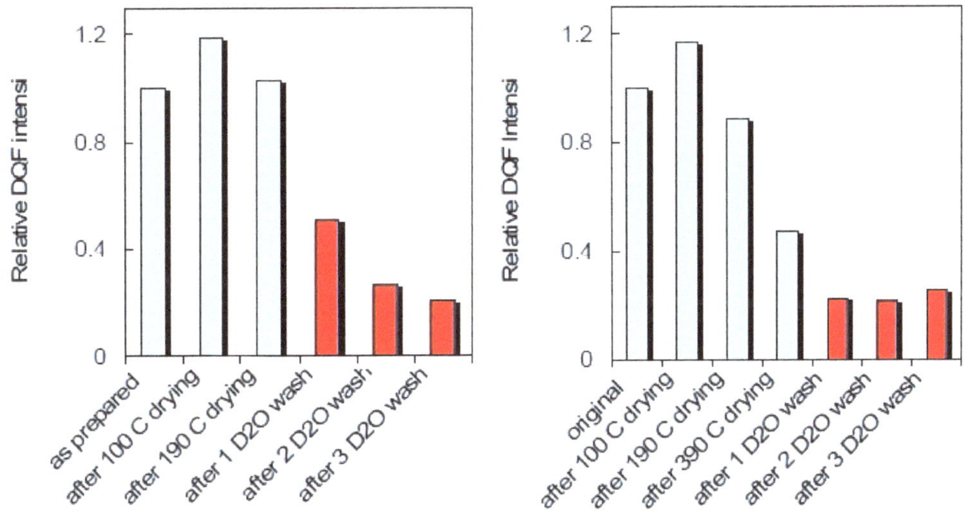

Fig. (21). Relative DQF intensities in the DQF NMR spectra for hydrated white cement samples, which were heating (oven) and washing (in D_2O) according to different programs of heating-washing actions. The NMR measurements (400 MHz) were carried out at 23℃.

Thus, a DQF signal of circa 17-18% of maximum cannot be removed, suggesting that it is from chemically bound water, *i.e.*, $Ca(OH)_2$, which is not destroyed by this treatment. Hence, the major part of the DQF signal in hydrated cements is from exchangeable water in the gel pores.

CONCLUDING REMARKS

^1H DQF NMR exploration of water in cement pastes discovered that a DQF NMR signal is sensitive to rigid, anisotropic water presented in hydrated cements. The study has also shown that the build-up of the proton DQF signal with curing time and the loss of the signal with sample drying are both consistent with expectation and the results of single-quantum relaxation experiments. The DQF spectrum line could be decomposed into at least two parts: the one is from protons of $Ca(OH)_2$, and other part stems from the water molecules of the planar C–S–H gel pores. The DQF line of C-S-H is broader than it could be normally expected for water, and that is considered as the effect of Fe^{3+} impurities. The DQF signal from the planar C–S–H gel pores could be removed effectively in the experiments with multiple D_2O washes. The deduced cut-off radius of 30 Å has presented the distance consistent with measurements of the surface density of Fe^{3+}. At fitting the grey cements data, the third DQF component of anti-phase Lorentzian doublet was discovered. This was associated with the water protons in the phases enriched by a significant C_4AF content.

ACKNOWLEDGEMENTS

The author thanks Andrea Valori and Peter McDonald.

REFERENCES

[1] R.J. Pellenq, A. Kushima, R. Shahsavari, K.J. Van Vliet, M.J. Buehler, S. Yip, and F-J. Ulm, "A realistic molecular model of cement hydrates", *Proc. Natl. Acad. Sci. USA,* vol. 106, no. 38, pp. 16102-16107, 2009.
[http://dx.doi.org/10.1073/pnas.0902180106] [PMID: 19805265]

[2] V.N. Izmailova, V.V. Rodin, E.D. Shchukin, G.P. Yampolskaya, P.V. Nuss, A.N. Ovchinnikov, and Z.D. Tulovskaya, NMR-approach in the study of the influence of nonionic surfactant on water state during cement hardening.*Nuclear Magnetic Resonance Spectroscopy of Cement-Based Materials.,* P. Colombet, A-R. Grimmer, H. Zanni, P. Sozanni, Eds., Springer-Verlag-Press, 1998, pp. 157-165.
[http://dx.doi.org/10.1007/978-3-642-80432-8_28]

[3] I. Maruyama, N. Sakamoto, K. Matsui, and G. Igarashi, "Microstructural changes in white Portland cement paste under the first drying process evaluated by WAXS, SAXS, and USAXS", *Cement Concr. Res.,* vol. 91, pp. 24-32, 2017.
[http://dx.doi.org/10.1016/j.cemconres.2016.10.002]

[4] L. Monteilhet, J.P. Korb, J. Mitchell, and P.J. McDonald, "Observation of exchange of micropore water in cement pastes by two-dimensional T_2-T_2 nuclear magnetic resonance relaxometry", *Phys. Rev. E.,* vol. 74, no. 6, p. 061404, 2006.
[http://dx.doi.org/10.1103/PhysRevE.74.061404] [PMID: 17280070]

[5] A. Valori, *"Characterization of cementitious materials by ^1H NMR",* PhD Thesis: University of Surrey, 2010.

[6] V.V. Rodin, P.J. McDonald, and S. Zamani, "A nuclear magnetic resonance pulsed field gradient study of self-diffusion of water in hydrated cement pastes", *Diffus. Fundam.,* vol. 18, no. 3, pp. 1-7, 2013.

[7] A. Valori, V.V. Rodin, and P.J. McDonald, "On the interpretation of ^{1}H 2-dimensional NMR relaxation exchange spectra in cements: is there exchange between pores with two characteristic sizes or Fe^{3+} concentrations?", *Cement Concr. Res.,* vol. 40, no. 9, pp. 1375-1377, 2010.
[http://dx.doi.org/10.1016/j.cemconres.2010.03.022]

[8] P.J. McDonald, V.V. Rodin, and A. Valori, "Characterisation of intra- and inter- crystallite C-S-H gel pore water in white cement based on an analysis of NMR signal amplitudes as a function of water content", *Cement Concr. Res.,* vol. 40, no. 12, pp. 1656-1663, 2010.
[http://dx.doi.org/10.1016/j.cemconres.2010.08.003]

[9] V.N. Izmailova, V.V. Rodin, E.D. Shchukin, G.P. Yampolskaya, and P.V. Nuss, "The effect of nonionic surfactants on the state of water in cement systems (by NMR relaxation data). 2. A model of pore space", *Colloid J.,* vol. 60, no. 1, pp. 13-18, 1998.

[10] J. Stepišnik, and I. Ardelean, "Usage of internal magnetic fields to study the early hydration process of cement paste by MGSE method", *J. Magn. Reson.,* vol. 272, pp. 100-107, 2016.
[http://dx.doi.org/10.1016/j.jmr.2016.09.013] [PMID: 27673276]

[11] V.N. Izmailova, V.V. Rodin, E.D. Shchukin, G.P. Yampolskaya, and P.V. Nuss, "The effect of nonionic surfactants on the state of water in cement systems (by NMR relaxation data): 1. The state of water in the course of structure formation", *Colloid J.,* vol. 60, no. 1, pp. 5-12, 1998.

[12] J. Dolinšek, T. Apih, G. Lahajnar, and R. Blinc, "Two-dimensional nuclear magnetic resonance study of a hydrated porous medium: An application to white cement", *J. Appl. Phys.,* vol. 83, no. 7, pp. 3535-3540, 1998.
[http://dx.doi.org/10.1063/1.366567]

[13] N. Bloembergen, E.M. Purcell, and R.V. Pound, "Relaxation effects in nuclear magnetic resonance absorption", *Phys. Rev.*, vol. 73, no. 7, pp. 679-712, 1948. [http://dx.doi.org/10.1103/PhysRev.73.679]

[14] J-Y. Jehng, D.T. Sprague, and W.P. Halperin, "Pore structure of hydrating cement paste by magnetic resonance relaxation analysis and freezing", *Magn. Reson. Imaging,* vol. 14, no. 7-8, pp. 785-791, 1996. [http://dx.doi.org/10.1016/S0730-725X(96)00164-6] [PMID: 8970082]

[15] F. D'Orazio, S. Bhattacharja, W.P. Halperin, K. Eguchi, and T. Mizusaki, "Molecular diffusion and nuclear-magnetic-resonance relaxation of water in unsaturated porous silica glass", *Phys. Rev. B Condens. Matter,* vol. 42, no. 16, pp. 9810-9818, 1990. [http://dx.doi.org/10.1103/PhysRevB.42.9810] [PMID: 9995232]

[16] H. Peemoeller, R.K. Shenoy, and M.M. Pintar, "Two-dimensional nmr time evolution correlation spectroscopy in wet lysozyme", *J. Magn. Reson.,* vol. 45, no. 2, pp. 193-204, 1981. [http://dx.doi.org/10.1016/0022-2364(81)90116-5]

[17] Y-Q. Song, H. Cho, T. Hopper, A.E. Pomerantz, and P.Z. Sun, "Magnetic resonance in porous media: recent progress", *J. Chem. Phys.,* vol. 128, no. 5, pp. 052212-12, 2008. [http://dx.doi.org/10.1063/1.2833581] [PMID: 18266417]

[18] V.V. Rodin, "Methods of Magnetic Resonance in Studying Natural Biomaterials", In: *Encyclopedia of Physical Organic Chemistry,* Zerong Wang, Ed., Chapter 53. vol. 4, part 4. John Wiley & Sons, Inc., 2017, pp. 2861-2908.

[19] V.V. Rodin, A. Valori, and P.J. McDonald, "Pore-water interactions in hydrated cement pastes by NMR", *Proceedings of the 9th international Bologna conference on Magnetic Resonance in Porous Media*, 13-17 July 2008, Cambridge, MA, USA, AIP Conference Proceedings.

[20] K.S. Mendelson, W.P. Halperin, J-Y. Jehng, and Y-Q. Song, "Surface magnetic relaxation in cement pastes", *Magn. Reson. Imaging,* vol. 12, no. 2, pp. 207-208, 1994. [http://dx.doi.org/10.1016/0730-725X(94)91517-2] [PMID: 8170300]

[21] H.C. Gran, and E.W. Hansent, "Exchange rates of ethanol with water in water-saturated cement pastes probed by NMR", *Adv. Cement Base. Mater.,* vol. 8, pp. 108-117, 1998. [http://dx.doi.org/10.1016/S1065-7355(98)00013-3]

[22] E.W. Hansent, H.C. Gran, and E. Johannessen, "Diffusion of water in cement paste probed by isotopic exchange experiments and PFG NMR", *Microporous Mesoporous Mater.,* vol. 78, pp. 43-52, 2005. [http://dx.doi.org/10.1016/j.micromeso.2004.09.015]

[23] H.M. McConnell, "Reaction rates by nuclear magnetic resonance", *J. Chem. Phys.,* vol. 28, no. 3, pp. 430-431, 1958. [http://dx.doi.org/10.1063/1.1744152]

[24] R.F. Feldman, and P.J. Sereda, "A new model for hydrated Portland cement and its practical implications", *Engin. J. Canada,* vol. 53, no. 8-9, pp. 53-59, 1970.

[25] V.V. Rodin, A. Valori, and P.J. McDonald, "A ^{1}H double quantum filtered NMR study of water in cement pastes", *New J. Physics,* vol. 13, p. 035017 (14 app), 2011.

[26] A. Valori, V.V. Rodin, and P.J. McDonald, "Spatially resolved & double quantum filter ^{1}H NMR studies of cementitious materials", *Proceedings of the 10th International Conference on Magnetic Resonance Microscopy,* p. 12, *The Heidelberg Conference*, 2009.

[27] V.V. Rodin, and D.P. Knight, "Molecular mobility of water in natural polymers: silk *Bombyx mori* silk with a low water content as studied by ^{1}H DQF NMR", *Biophysics,* vol. 49, no. 5, pp. 730-737, 2004. [PMID: 15526463]

[28] G. Navon, U. Eliav, D.E. Demco, and B. Blümich, "Study of order and dynamic processes in tendon

by NMR and MRI", *J. Magn. Reson. Imaging,* vol. 25, no. 2, pp. 362-380, 2007. [http://dx.doi.org/10.1002/jmri.20856] [PMID: 17260401]

[29] U. Eliav, M. Komlosh, P.J. Basser, and G. Navon, "Characterization and mapping of dipolar interactions within macromolecules in tissues using a combination of DQF, MT and UTE MRI", *NMR Biomed.,* vol. 25, no. 10, pp. 1152-1159, 2012. [http://dx.doi.org/10.1002/nbm.2784] [PMID: 22362643]

[30] U. Eliav, and G. Navon, "A study of dipolar interactions and dynamic processes of water molecules in tendon by ^{1}H and ^{2}H homonuclear and heteronuclear multiple-quantum-filtered NMR spectroscopy", *J. Magn. Reson.,* vol. 137, no. 2, pp. 295-310, 1999. [http://dx.doi.org/10.1006/jmre.1998.1681] [PMID: 10089163]

[31] V.V. Rodin, L. Foucat, and J.P. Renou, "The dipolar interactions and dynamics of water molecules in collagen fibres from two connective tissues of different ages by ^{1}H double-quantum-filtered NMR spectroscopy", *Mater. Sci.,* no. 5, pp. 2-10, 2004.

[32] A. Bax, R. Freeman, and S.P. Kempsell, "Natural abundance ^{13}C – ^{13}C coupling observed *via* double-quantum coherence", *J. Am. Chem. Soc.,* vol. 102, pp. 4849-4851, 1980. [http://dx.doi.org/10.1021/ja00534a056]

[33] R. Kimmich, *NMR Tomography, Diffusometry, Relaxometry.* Springer: Berlin, 1997.

[34] W. Derbyshire, M. Van Den Bosch, D. Van Dusschoten, W. MacNaughtan, I.A. Farhat, M.A. Hemminga, and J.R. Mitchell, "Fitting of the beat pattern observed in NMR free-induction decay signals of concentrated carbohydrate-water solutions", *J. Magn. Reson.,* vol. 168, no. 2, pp. 278-283, 2004. [http://dx.doi.org/10.1016/j.jmr.2004.03.013] [PMID: 15140438]

[35] G.E. Pake, "Nuclear resonance absorption in hydrated crystals: fine structure of the proton line", *J. Chem. Phys.,* vol. 16, pp. 327-335, 1948. [http://dx.doi.org/10.1063/1.1746878]

[36] G.D. van Velzen, and W.Th. Wenckebach, "Nuclear magnetic ordering in $Ca(OH)_2$. I. Molecular field approximation", *J. Phys.,* vol. 51, no. 13, pp. 1445-1461, 1990. [http://dx.doi.org/10.1051/jphys:0199000510130144500]

[37] C. Badea, A. Pop, C. Mattea, S. Stapf, and I. Ardelean, "The effect of curing temperature on early hydration of grey cement *via* fast field cycling-NMR relaxometry", *Appl. Magn. Reson.,* vol. 45, p. 1299, 2014. [http://dx.doi.org/10.1007/s00723-014-0565-z]

[38] P.S. Wang, M.M. Ferguson, G. Eng, D.P. Bentz, C.F. Ferraris, and J.R. Clifton, " ^{1}H nuclear magnetic resonance characterization of Portland cement: molecular diffusion of water studied by spin relaxation and relaxation time-weighted imaging", *J. Mater. Sci.,* vol. 33, pp. 3065-3071, 1998. [http://dx.doi.org/10.1023/A:1004331403418]

[39] P.J. McDonald, V.V. Rodin, and A. Valori, "Microstructure and pore-water interactions in cement based materials", *14th Full Meeting of the NANOCEM CONSORTIUM, October 23–25,* 2007, p. 226 Madrid (Spain).

[40] V.V. Rodin, A. Valori, and P.J. McDonald, "^{1}H DQF study of cement pastes. In:", *Abstract Book "Frontiers of Magnetic Resonance", BRSG: The Magnetic Resonance Group of the Institute of Physics, University of Warwick (UK),* p. 56, 2008.

[41] V.V. Rodin, A. Valori, and P.J. McDonald, "How ^{1}H DQF NMR spectroscopy highlights water in hydrated cement based materials", *Abstract Book: "The Fred Glasser Cement Science Symposium", University of Aberdeen (UK),* p. 77, 2009.

[42] Y. Liu, A. Gajewicz, V.V. Rodin, J. Keddie, and P.J. McDonald, "Explanations for water whitening in secondary dispersion and emulsion polymer films", *J. Polym. Sci., B, Polym. Phys.,* vol. 54, no. 16, pp. 1658-1674, 2016.

[http://dx.doi.org/10.1002/polb.24070]

[43] S.F. Tanner, B.P. Hill, and R. Parker, "Interactions of sorbed water with starch studied using proton nuclear magnetic resonance spectroscopy", *J. Chem. Soc., Faraday Trans.,* vol. 87, no. 16, pp. 2613-2621, 1991.
[http://dx.doi.org/10.1039/ft9918702613]

[44] J.G. Powles, and J.H. Strange, "Zero time resolution nuclear magnetic resonance transients in solids", *Proc. Phys. Soc.,* vol. 82, pp. 6-15, 1963.
[http://dx.doi.org/10.1088/0370-1328/82/1/303]

[45] W.R. Busing, and H.A. Levy, "Neutron diffraction study of calcium hydroxide", *J. Chem. Phys.,* vol. 26, pp. 563-567, 1957.
[http://dx.doi.org/10.1063/1.1743345]

[46] J-S. Lee, and A.K. Khitrin, "Accurate measurement of ^{13}C-^{15}N distances with solid-state NMR", *J. Chem. Phys.,* vol. 124, no. 14, p. 144508, 2006.
[http://dx.doi.org/10.1063/1.2191043] [PMID: 16626215]

[47] H.F. Taylor, *Cement Chemistry.* 2nd ed. Thomas Telford Publishing: London, 1997.
[http://dx.doi.org/10.1680/cc.25929]

SUBJECT INDEX

A

B

D

E

F

G

H

I

L

M

N

O

P

Q

R

S

T

U

V

W

www.ingramcontent.com/pod-product-compliance
Lightning Source LLC
LaVergne TN
LVHW070950210926
848482LV00002B/159
* 9 7 8 1 6 8 1 0 8 6 3 0 9 *